D1093156
PLETTENBERG BAY
Plettenbergbay
IN
Where the Black Eagle flies
BRAKKLOOF
ROBBERG
Jack's Point
OCEAN

BLACK EAGLE FLY FREE

Dear Ian,
Happy Christmas
love from
Chris & Jean.

Parent and eaglet (Photo: W. T. Miller, See overleaf)

Black Eagle Fly Free

J. A. Cottrell

PURNELL

CAPE TOWN · JOHANNESBURG · LONDON

PUBLISHED BY PURNELL AND SONS S.A. (PTY) LTD.

KEEROM STREET, CAPE TOWN

SBN 360 00100 9

SET IN 11 ON 13 POINT BASKERVILLE

PRINTED AND BOUND IN SOUTH AFRICA BY
GOTHIC PRINTING CO. LTD., OBSERVATORY, CAPE TOWN

Dedication

In the service of wildlife protection
For those in peril in our land

and for
CHRISTOPHER
who knows more of
Nature's secrets

CONTENTS

YOUR GUIDE

This is a true story and all its characters and places are real. There is only one exception: Eagle Hill, round which the story centres, is not the true name, the name given on the maps, of that stony promontory on which the voice of the winds is seldom stilled. It dominates a wildlife sanctuary where our heritage is being conserved. It is also a private estate and the home of its owner. My duty to it is my conscience, for I am its warden. I must keep it inviolate. With its real name sacrosanct there will be no betrayal.

The hill itself and the wild stretch of coastal cliffs surrounding it have been faithfully described. They are on the southern coast of the Cape Province in South Africa.

I am happy to be your guide on a nature trail which will lead us to share, as well as a man can do, the life of a Black Eagle and his triumphs equally with his disasters.

One of the Black Eagles we shall see is Ave, the male or tercel, also known as Verreaux's Eagle. He and his family will be the focus of our study, but we will also meet other birds of prey and find that Black Eagles are distinguished members of the eagle-kind. You might think that Ave's name came from his tribe's scientific name, *Aquila VErreauxi*, but in fact he is also a legendary character with the mythical name Ave Avatar. As your dictionary will tell you, Ave means hail and farewell, and Avatar is in Hindu legend an incarnation of a god of the skies.

Ave's mate is named Mavi, Princess of Dragon. The eaglet born in 1968 is named Eaglet. Those born earlier are called Sixtyfive and Sixtyseven, after the year of their birth. This story tells of Eaglet's life and upbringing, but the others are mentioned from time to time. Black Eagles attain maturity after four years and, when living free, have a lifespan as long as that of man.

Eagles do not need names. There is no need to humanise them. But I have named mine for the sake of clarity and, I must confess, because of the delight their company has given me.

Russet the Red Setter was my sporting dog who served my sons and me in our shooting days in the wilds. Later he became my ranger in the Eagle Hill nature reserve. The staunchest of friends to all his fellow creatures, he had an amazing understanding of the ways of the wild and I learnt much from him.

When I wrote this story it was contained in a longer natural history manuscript covering a wider field. It was Siegfried Stander who extracted and reshaped it and I am deeply grateful.

I have relied on my own observations and general background knowledge, acquired during sixty years of close contact with Nature. I believe too that if the study of natural history is to endure we must take it out of the lecture room and return it to the field; we must concentrate less on the bare bones and retain something of the romance and adventure.

The photographs reproduced in this book have been contributed by my naturalist friends. The reason for their generosity is easily explained. They are all wildlife protection crusaders. Among them is W. T. Miller, a great master of wildlife photography.

As you will have gathered, my detailed study of the Black Eagles would not have been possible without the co-operation of Eagle Hill's owner and benefactor. It was he who allowed me to build the observation hides and who helped in many other ways. Although I may not name him, for reasons known to you, my thanks are due and sincere.

JOHN COTTRELL

Plettenberg Bay,
June, 1969

CHAPTER ONE

Ave returns

And then I saw it. No more than a tiny speck in the sky, smaller even than the dot over the "i" in "tiny", it appeared to be stationary over the high mountain peak. It was black.

I knew at once that this could be what I had been searching for. I watched it intently, not daring to move my field glasses and take my eye off it for one second for fear of losing it. It was definitely a bird. The brilliant light thrown over the mountains by the westering sun, now behind me, would reflect or obliterate shimmering aircraft metal at this distance. If the bird was approaching, as it seemed to be doing, I knew that only an eagle-falcon, or lammergeyer or vulture could ride the wind in such a dead straight line. They travel like high velocity bullets, throwing neither up nor down, right nor left, slap on target.

Of these three, only one was black, pitch black in front—the eagle. And its swift glide was so smooth that one could imagine it soaring down the long lance of an individual shaft of sunlight. Moreover, lammergeyers and vultures no longer fly the skies of the Southern Cape coastal belt.

So, I concluded, it was my friend Ave, the Black Eagle, whose home was on Eagle Hill's sea cliff behind me.

There could be little doubt about it now. He was riding the

right sun ray and already well past his long range marker, Formosa Peak, which he had probably used before on this annual journey from the interior.

Flying so high, he would be able to see far in every direction and take in, without turning his head, all in his wide ken. His powerful eyes focus independently of each other, presenting two separate images for him to interpret simultaneously. He is able to swivel his eyes round in all directions and obtain front and back views without deflecting from his course.

Soaring over the coastal uplands, Ave would have in panoramic view wide beaches, and the long line of surf. He would scan the deep gorge of the Keurbooms River and the long estuarine lagoon into which it and the Bitou River flow to mingle with the tidal waters of Plettenberg Bay. To his left he would survey Robbe Berg Cape with its bald sand-patch nape and cliff faces. He would see the lighthouse on the bluff above and Seal Point below. Ahead he would notice Beacon Island across the Piesang River lagoon and the village on the hill.

Although these coastal skies are radiantly clear, because the prevailing winds come from the southern oceans and are dust free, it would hardly be possible for a human observer on Eagle Hill, however well-trained his eyes might be, to see an eagle directly over Formosa Peak. Perhaps I had not, in fact, first sighted him there. But so high and true was his course that it appeared to be so. On the other hand, reflected light from the water below may have helped vision. Wherever he was, time stopped when I first spotted him.

I had flown over the intervening space in a light aircraft so I could imagine him now surveying our fresh water dams —glittering jewels when seen from the air—which were the very heart of the wildlife sanctuary developing round Eagle Hill. Now the valley streams, with their deep weir pools, would appear as well-fed snakes wriggling their way to the sea through colourful patches of wild-flowers established where a jungle of alien vegetation once stood.

Now I dropped my glasses and saw again, as an enduring miracle, the perfection attained by this graceful, nicely-poised, fast-gliding eagle. Ave swung gently upwards at last against the light sou'wester. I could see his yellow face and feet and his black fully-feathered legs lying like splines under his longish tail and working in unison with it in flight. And then he was almost overhead with his long obverse wings slightly raised and fanned quills upswept, as he smoothly reduced speed. When he swung over me he held wind with a few quick beats and dropped his feet, claws clenched in greeting. Then he banked so that I fancied I could see the sunlight in the western sky through the windows in the broad, white inner webs of his flight feathers.

I imagined, then, the start of Ave's long journey earlier that day, in the high Maluti Mountains of Lesotho, 600 miles away to the north-east, and my thoughts followed him through the earlier stages of his flight from the interior.

In my Lesotho days, I saw Black Eagles set out on such journeys and observed how they used winds and currents. When trekking from valley to valley in the Malutis, it had been my habit to make for the top of a pass or neck before noon and there off-saddle on a watershed ridge where the horses would be hobbled and put out to graze. The long midday break gave opportunity for watching the movements of Black Eagle families, lammergeyers and other birds of prey.

Horsemen who ride this great tumbled mass of windswept mountains know the strength of the air currents and change-winds that trade there, rustling and whirling up valley and kloof and mountain and skywards, when, spent and dispersed, they gather later to drop down again through cloud and cool mountain air and, freshening, go frisking through the shadowed land. There was a time when I pitched my light bivouac tent in a mountain valley only to have it flattened by a rushing wind passing up the valley at dusk. An hour later, when the tent had been put up and scattered belongings recovered, the wind came tearing down the valley once again. Then all was still for the night.

The malicious mischief of whirlwinds in the rising heat of the day is too well known for elaborate description. One of these dust-devils nearly killed me by lifting the iron roof of my office high into the air, to slam it down on my doorstep as I was about to rush out after recovering from the shower of brickbats.

Although I had come to Eagle Hill early in order to make sure that I would not miss seeing Ave fly in, I knew that he would not have made an early start because that would have been to little avail. He would have waited until the mountains and their deep intersecting valleys were well warmed by the hot sun: until their rising air created up-spiral currents, like chimneys pouring and rolling their smoke up into the clouds.

Then Ave had launched himself with a few wing beats, slipped over the ridge and made for an eagle airport in one of the broad valleys which drop down from the highest watersheds. Seldom flapping a wing now, except to bank or change currents, he would have flown up the valley, adroitly exploring every current from every direction, sailing up like a kite, feeling and probing the air all the time with his sensitive wing and tail feather indicators, until eventually he had struck an air spiral of near dust-devil force. In this he would have risen almost vertically, circling rapidly round and round in small revolutions, like an ant crawling round the thread of a screw, and still without a stroke of the wings. At an altitude determined by the now spent centrifugal force of his whirling dust-devil Ave's own turning circles would have widened out and here, perhaps six or seven thousand feet above the nine or ten thousand feet high Maluti, the planing eagle would take over direction control and head for home on the second stage of his journey.

This first stage had been much the shorter and harder one for Ave. For perhaps twenty minutes every muscle, every flight-feather and every nerve in his tough body had been strained and every innate and nurtured skill fully employed; yet he was perhaps further away from his journey's end than he was at his starting point below, because the spiral may have borne him northwards to a point above the highest peaks of the Natal Drakensberg and the sources of the Tugela and Orange Rivers. This would happen if the high altitude wind was in that direction.

And yet this was the very wind Ave had expected and planned to use for his six-hundred-mile speed glide down to sea level. Moreover, he had timed the start of his journey so that he would be poised at optimum height at midday. By then the cool air from the southern seas would be at work, creating the strong up-current conditions most favourable for speedy flight. The heavier the eagle relative to its size the faster its rate of descent and capacity for speed. The bird uses the momentum of its falling force against the current in the selected airstream to drive itself forward, skilfully employing its banks of flight feathers to ride the air flow. In favourable conditions it planes along, gaining incredible speed, on a long steady glide to the sea coast.

It has been estimated by experienced glider and light aircraft pilots that the Black Eagle tercel, with an estimated falling rate of four feet per second, is capable of sailing at a speed of 120 miles per hour on a flight of this nature in favourable air conditions. That being granted, a journey of 600 miles would take about five hours. The feat is the more remarkable because throughout the journey the Black Eagle would have cruised with hardly a wing-stroke except to launch himself from his rock perch on the mountain crag at the starting point and to find an up-draught spiral; and perhaps to break speed in the last few yards as he came up to perch on the sea cliff.

Having completed the first stage, Ave had instinctively selected the south-west air stream and followed it unerringly on his homeward bound journey. But there can be no doubt whatsoever, in my view, that his truly wonderful eyesight is also fully employed, watching for the familiar landmarks as he soars along at great height. With the Maluti Mountains behind him and Aliwal North ahead, he crossed the Lesotho border noting the confluence of the Orange and Caledon Rivers and, incidentally, passing over the region of the Karroo Red Beds where fossilised bones of his prehistoric ancestors were found.

Then the Compassberg, Toverberg, Spandaukop and other sentinels of the Karroo came into view.

Now high over the Cradock Mountain Zebra National Park and about 200 miles from home, his far-seeing eyes scanned the azure skies above the southern escarpment ranges, seeking out Formosa Peak where she stands bare-footed by the sea, holding her head erect in the sunlight. Ave, the Black Eagle, on his home-coming journey had probably followed this known way annually for close on forty years.

After passing Formosa, Ave would have spotted me on Eagle Hill from afar. Although I was wearing bush clothes which blended perfectly with the stony background and Russet, lying in my shadow on the red-lichen covered rocks, would be quite invisible to an airman, Ave's penetrating eyes would never fail to pick us out. He spots, with remarkable ease, hares in their forms and dassies taking cover in vegetation, rocks and scree or in dark crevices.

He came soaring overhead and we exchanged greetings. Then, Ave dived over the cliff-face, still moving into the west wind, and wheeled sharply round over the sea, showing the white patch covering his back and rump. He came back flying eastwards with the wind now, then along the crescent shaped cliff, passing his two nests, before he swung up to settle on his favourite rock perch. And there he rested with the serene composure these eagles find immediately a journey is over or task completed.

Ave was home. It was the 6th February, 1968.

CHAPTER TWO

In search of Nature

A grey Land Rover in four-wheel drive climbed steadily up the narrow, stony track, its bulldog nose pressing forward, its square flanks sometimes brushing the five-foot high fynbos scrub. It reached a level terrace, where scrub gave way to purple heath. Then a young Grysbuck doe scurried across the track, her sleek rufous body bright in the sunshine. Seconds later her broad inquisitive face, with large oval ears alert and black nose sniffing the air, peered through the heath as she watched the Land Rover pass by. A moment later, fearlessly, she stepped back into the track and went on down to the valley to browse in the wildflower garden.

She knew the Land Rover and both its occupants and they were friends. She had scented and recognised them. She would not heed the warnings of other buck to dive into a secret bush tunnel and disappear whenever man or dog appeared because she had learned to distinguish friend from enemy. And the man and the dog in the Land Rover knew her too. She was Runi who, a year before, had narrowly avoided being run over by a car when, terror stricken, she broke across a road to escape from a pack of marauding dogs. She had been picked up, fortunately before the dogs could tear her to pieces, but dazed and with a badly injured shoulder: then nursed back to health in a dark, straw-padded garage. Finally she had been brought to her new wildlife home where jackal-proof fencing kept her safe and where a Red Setter befriended her.

At the end of the track the man swung the Land Rover broadside across the grassy slope of the turning circle and brought it to a stop. He stepped out, walked round and opened the far door for Russet, his old gun dog, now rising ten. Together they climbed the final slope to their look-out post on the summit of Eagle Hill. Turning towards the interior the man searched the skies above the distant mountains. Satisfied that nothing was in view there he dropped his eyes to survey the wildlife sanctuary immediately below. Then the immediate scene faded as thoughts of the past came flooding in.

This high hill on the coastal cliffs had been used as an observation post for patrol work ever since the discovery of its ancient secret. Not only does it offer a clear view of the coastal belt, through which the scenic Garden Route runs today, and of the mountain ranges beyond, but from it alone can be seen the waters of both Still Bay in the west and Plettenberg Bay across the peninsula in the north-east and their sea approaches from all sides. No ship could pass along these shores or enter either bay without being observed by watchers on Eagle Hill. And here on its summit lies hidden the evidence of a long-forgotten story of the past.

In the hill-top scrub stands a sturdy cone, built of loose stones, at the apex of which is fashioned a rustic seat and, behind it, a stone beacon such as ship-wrecked sailors used for supporting their distress signal staff. Perhaps four hundred years or more ago Portuguese sailors, exploring these wild coasts between the Cape of Storms and Storms River Mouth, had attempted a landing and lost their ship or maybe later a Dutch East Indiaman had been wrecked on a submerged reef. Perhaps the castaways had anxiously kept watch here for a ship to rescue them from a hostile land and its aggressive aborigines.

Only two centuries ago this land of Formosa—"beautiful" was the name the early Portuguese navigators gave the Bay and its surrounding mountains—was the hunting ground of Bushmen and Strandlopers who lived in the caves along the coast. But for many thousands of years before it was the home of stone age peoples of the early, middle and late periods, whose artefacts still lie scattered everywhere in the caves, river beds and hills. The skeleton of a child who lived about 2,700 years ago was recently discovered in the upper midden of an archaeological working in a cave nearby. These primitive peoples lived here because game abounded in this region of regular rainfall. For countless years the South Western Cape had been the stamping ground of elephant, hippopotamus, buffalo, antelopes, bushbuck and smaller game. Leopard, lion, lynx and lesser predators hunted them. Birds of brilliant green and red plumage—loeries, parrots and trogons—winged through the magnificent evergreen forests and clear-winged dragonflies, damselflies and brightly coloured butterflies darted along broad valleys and mountain streams. And from the earliest times Black Eagles had kept watch over Eagle Hill.

Into this wild and beautiful land of primeval forests, fertile valleys and mountains came a few white explorers and hunters, followed by scattered settlers. Then, in the seventeen-seventies, exploitation of the indigenous forests by the Dutch East India Company began and ships sailed into Bahia Formosa—soon to be renamed Plettenberg Bay by the Dutch Governor, Joachim van Plettenberg—to carry the timber away. The Black Eagles would have been among the first of the land birds to see these winged ships. Perhaps they felt uneasy when the ships sailed into the bay. Their domain was being invaded. Perhaps they were warned by instinctive fear of the unknown, or by the memory of shipwrecked sailors on Eagle Hill. Perhaps an ancestral memory, which had served their tribe so well through its long history of survival, would have told them of the disruption which would follow these messengers from a far shore. Inevitably, when giant two-thousand-year-old yellow-wood, stinkwood, Cape Ash and other great trees were felled, the animals and birds were driven off. Those which fed on trees, vines and shrubs destroyed in timber operations could never return until the food-bearing flora were re-established.

Civilisation pressed ruthlessly on and wildlife gave way before it. Towns and cities grew where the wilderness had been. And eventually the cycle was full and civilised man in the cities began to yearn again for the life of the wilds, for the forests and the open veld, the rivers and the seashores. But now his yearning held compassion for the wild things that still survived and he reached out his hand in friendship. Deep in man his true spirit had quickened.

In such a way it happened that a private nature reserve was established round Eagle Hill and the work of attracting back, re-introducing and protecting indigenous wildlife was begun with the aid of the Cape of Good Hope Provincial Administration's Department of Nature Conservation whose sound practical advice, based on long experience and research, was made freely available. One example of this will suffice. A list was supplied of indigenous trees and shrubs upon which birds and buck, once known to have lived here, fed before axe and fire destroyed them, and quick growing alien vegetation suppressed regrowth. Years of research on local flora and fauna went into the preparation of that list.

It is essential to ensure adequate protection from poachers and roving dogs for game rehabilitated in areas where uncontrolled hunting was rife in the past. Wildflower protection is no less urgent. Men and money on a prohibitive scale would be needed for this job. Most realistically, therefore, the Director of Nature Conservation has enlisted a corps of Honorary Nature Conservation Officers. In the Cape Province last year there were 551 Honorary Nature Conservation Officers, including 236 Officers and Inspectors of Administrative departments whose duties involve travelling, and 315 volunteers. Among the latter are conservation-minded farmers who practise what they preach. Some of them have preserved game on their farms for many generations, shooting with restraint and less freely than in the earlier days when it was plentiful. Those fondest of animals and their natural surroundings are often the keenest sportsmen and best shots.

I have been a hunter too, for hunting does not mean just shooting for the pot if one is a sportsman. Studying the habits of game and learning the arts of spooring and tracking provide healthy recreation in the bush.

Two Lenje tribesmen were always available for my hunting trips in the Central Province of Zambia. One named Longwani, who was very tall and as straight as a ram-rod, was exceptionally good at spotting distant game in open or park-like country. The other, Lioni, was short and stocky and no fresh spoor, or warthog hiding behind a log, ever escaped his sharp eyes. These men had an intimate knowlege of the country and of the movements and habits of the game. They were expert trackers. They were also very observant of bird life. With such excellent bush-craftsmen to guide me, hunting and ornithology could be combined and enjoyed to the full.

The shooting of rare game or taking easy shots at elephant, giraffe, hippopotamus and rhinoceros had no appeal whatsoever when there was plenty of adventure to be had stalking the dangerous buffalo bull or sable bull guarding the rear of a big herd. To stalk a lone hartebeest bull, already on the alert, round a bush-fringed dambo for a couple of hours and hold fire until one could see the dark chest-patch behind his shoulder, which one must aim at for the fatal heart-and-shoulder shot, called for hunting skills and a well-trained eye.

Then there were the game birds. Guinea-fowl, francolin, duck and others abounded and for them a dog was needed if one wanted really good sport. And so first Russet's uncle and then Russet himself were trained.

Russet's father was a pedigree Red Setter imported from Ireland and his mother a commoner but salted and hardened to tropical bush conditions. Russet was acquired at the age of one month and at seven months his gun-dog training started. He had a very good nose and made excellent progress because, from the start, it became evident that his one aim and purpose in life was to please his master. He loved his work and the excitement of seeking and finding and pointing and flushing. Then when the command "fetch out" came he obeyed, retrieving the bird and carrying it back ever so gently in his soft mouth, to be rewarded by his master's pleasure.

As he grew older it became more and more obvious that he did not like this last duty. Unhappiness, almost guilt, stole across his happy, friendly face as he dropped the dead bird into his master's bag. He would turn away quickly and pretend that he had had nothing to do with that side of the job. Praise when the day's bag was laid out at home meant nothing but sadness to him, so that ceremony was soon done away with. He would never eat game birds in any shape or form.

There came a day when we went for a walk in the forest reserve without the gun. An objective was to visit a Little Sparrowhawk's nest from which one of three young birds was taken later for training in falconry. This done, we were about to return when the bay of hunting dogs was heard. We ran over the ridge to see them chasing a baby duiker. Urged on, Russet bounded forward and caught up with the pack which fled in fear on seeing him.

When I came up Russet was gently licking the furry little animal, which had collapsed. After it had recovered we put it in some long grass and began to walk away but the fawn followed Russet and would not leave him. We waited until dusk when the parents, who had been watching, emerged from cover. Then we slipped away.

Russet and I never hunted game or shot birds again after that. A chapter in our lives, which we had enjoyed but were now ready to forget, had been closed.

At first the change perplexed Russet when I took him for walks carrying a stick instead of the gun. He would stand pointing at a bird he had marked, with one front paw raised and the silky feathers in his outstretched tail waving in the breeze, expecting the old routine, only to be called away. He simply could not understand why. Sometimes he would run away down wind and flush guinea-fowl right over my head and, when there was no report, he would come and look at me and the stick, his head on one side, wondering why we had failed in our duty. But I smiled, so all was well.

When we left the north, Russet stayed on for a time with friends who were kindness itself and he became very much the pet of the family. Later he came on by air to join us again.

He was now ready for his new training in the wildlife sanctuary. He quickly and gladly learned that his work was to find birds and mark them for me to come up and stand with him until the young birds in the covey came from cover, curious to see more of his waving, feathery tail. This enabled me to count the coveys. And so began a new and exciting friendship.

Russet is now an old dog with grey hairs round his muzzle

and the new life suits him well. He is my game ranger. We are old hands in the wildlife field and work as a pair, like Longwani and Lioni did. We are always on the look out for poachers. On occasion Russet has smelt out suspects concealed in the thickets and warned. Then I have challenged: "Come out or I will set the dog on you!" Afraid to run, the suspect would emerge. Then the man is closely questioned to find out exactly what he has been up to and where he has set his cruel, thin wire, noose snares that are the hangman's halter for innocent fleeing Grysbuck. Invariably these snares are set in the entrances to the very tunnels the buck themselves make through the scrub as a means of escape, principally from marauding underfed dogs, since jackals have been exterminated in the Southern Cape. The tunnels are cleverly designed with a maze in the middle from which smaller paths disperse in all directions and peter out, leading the pursuing predator astray in the thickets.

If the poacher suspect does not admit what he has been up to, I order him to accompany me to the Police Station, pointing to the Land Rover, and warn the man of the heavy sentences passed on people convicted of game trapping. Then I tell him that the sentence might be reduced if he disclosed where his snares were, because many a buck has been strangled and left to rot in an abandoned snare.

No warning, without prosecution, may be given by a conservation officer to anyone illegally destroying fauna or flora, including the picking of wildflowers. His duty is to prosecute, and rightly so, for we South Africans are determined to conserve our once fast disappearing wildlife heritage.

CHAPTER THREE

Friendship renewed

As Ave circled me on the day of his homecoming, with feet still down and wings inclined, the golden light sifted through his latticed wing windows. In that happy moment of recognition, there flashed across my memory a picture of a young Black Eagle circling feet down, 40 years ago in summer-clouded Maluti Mountain skies.

I had often seen this action over Eagle Hill in recent years and always studied it in wonder but with eyes so blind that I had not realised who Ave was. He remembered me through all these years yet I did not know him until this moment of miraculous revelation and joy transcending all common experience when an old friendship is renewed.

Ave and Mavi may well be aware that when we are about on Eagle Hill their enemies, such as the baboons, do not trouble them. That may be one explanation of their friendliness. But Ave has always been the more demonstrative of the two. From the earliest days of our association here at Eagle Hill he has greeted me, and sometimes come to perch within a few yards of me without any misgivings, when I am alone on the nesting cliff. These gestures have never failed to evoke wonder as well as give great pleasure.

I have wondered about them because I believe that Black

Eagles are quite incapable of affection for man. A mated couple do most definitely show, by human criteria, great affection for one another and their dependent young. One of the ways in which mating birds show it is to tumble in courtship play, with feet down and claws locked together. Ave's greetings for me are much simpler: just a drop of the feet as he swoops along the cliff or a slight hover and drop of the feet when he comes overhead. The wonderful thing about them is the seemingly spontaneous warmth; but that is my human reaction which, I still maintain, is not shared by Ave in his simple act of trustful recognition.

Black Eagles are sometimes more trusting of mankind than man deserves and they really do trust those they know well. There could be special reasons for an individual bird to salute a particular man, and an old friendship is one.

The reason for Ave's greetings was enveloped in mystery until he circled round us over Eagle Hill on that day. Then his action suddenly recalled memories of an old association. When on trek in the Maluti Mountains forty years ago, I found a juvenile, recently fledged, Black Eagle caught up in a low bush beside the bridle path and in serious trouble. A spring trap, with severed rawhide thong attached, clamped his left tarsus, which was lacerated but not broken, and he was suffering from exhaustion and starvation. I nursed him back to strength. I kept him in a Basuto basket which was secured above a pack-saddle when we travelled. Sometimes I carried the basket across my knees until the pack ponies settled down. I always rode in front and my horse orderly in the rear with the pack and spare ponies in single file between us.

After about a fortnight the young eagle recovered and was released. For several days he followed us, coming down for his dassies, which I shot for him and put out on his basket when we camped or rested at midday.

His action of circling round with feet down instead of stooping to take his dassies is vivid in my memory.

Ave, too, never carries with his left foot, as Mavi does. And on two occasions recently, when his own fledgling was in grave trouble, he came to me. I have no scientific way of telling if Ave is in fact that Black Eagle of forty years ago. A ring would have been the best evidence but the idea of ringing wild eagles has no appeal for me because of its dangers for the bird. In my own mind, however, I am sure. Birds have excellent memories and survival experience is strongly imprinted, so the eagle would be aware of the old association.

Friendships with wild birds and animals are for naturalists as precious as their discoveries. Free of the complications of human relationships, they are as sincere as honey without wax. Of all the words describing man's happy relations with wild creatures "friendship" is the loveliest and the best. Embracing trust and understanding, fondness or love, it gives meaning to our emotions. True friendship between man and wild animal is simply warm affection as between man and man or man and dog. Love of fellow men and fondness for fellow creatures are akin.

In all my days of natural history observation few were filled with richer experience than this day of Ave's return. To be enraptured by his superb presence would have been reward enough without being surrounded by the rare atmosphere of mystery and involvement. Such experiences serve to increase the capacity for wonderment which is so necessary to natural history research. The twin spirits of enquiry and wonder inspire discovery.

I knew better than to approach him there and then. I had welcomed him home with my usual gesture of palm up on outstretched arm. All was well and he should be left in peace because that was all he wanted. Seated now on Eagle Hill look-out post I was soon deep in thought.

During the last four years I have studied and tried to understand this Aquila verreauxi family's way of life whilst closely guarding their nest and interests here. My records of previous nesting seasons clearly indicated that I could expect the male bird, Ave, to return from summer quarters during this, the second week of February to defend his nesting territory, after a midsummer absence with his family of from three to six weeks duration. His mate, Mavi, follows ten or twelve days later, leaving their recently fledged eaglet behind.

Ave is smaller by almost one-third and more falcon-like than Mavi who is heavier and probably older. It is normal for the female Black Eagle to be appreciably larger than the male or tercel. The word tercel or tiercel is derived from the Latin word tertius, third, because Roman falconers believed that a hawk's third egg produced a small male. Scientific usage would probably not admit of my description of Ave as an eagle-falcon tercel. I see him as such.

Early on the morning of Ave's return I had a hunch that he would come home that day and arranged my day's work so that I would be able to spend the afternoon on Eagle Hill, after a patrol round the nature reserve with Russet. The day was fine and clear after rain and when the south-west breeze freshened my expectation increased.

February was the time of year when the mating urge first began stirring in Black Eagle blood: when they would have to leave their previous year's single off-spring in the area to which they had taken him or her late last year and spent the midsummer weeks there with him, giving him his final lessons in the intricate art of Black Eagle living and self preservation. They would probably never see him again or be part of his daily existence, for he would never be allowed to enter his parents' home territory again. The annual break-up of a very intimate family life is sad but inevitable. It is part of life's plan for Black Eagles. For the parents the pang of parting is probably softened by the strong compulsion of the renewed onset of the reproductive urge and the fact that parental-care instincts are on the wane. The seasonal wheel has turned full cycle.

For the young eagle the loss of his parents at the age of seven months would be hard or easy to bear according to his temperament. Every eaglet is born with his or her own individuality. What young animal is not? A strong desire for independence and feeling of self-reliance would make parting easy, particularly if companionship with another young eagle had ripened. In three year's time, when he was nearly four years old and fully mature, he would be ready to establish his own home. In these circumstances, it would be possible for the parents, each in turn, to slip away unnoticed before they made the long flight back to their eyrie together.

But if the youngster still depended on the family association, the separation would be hard for all and the parents would have the unpleasant duty of letting him know bluntly that companionship with them was no longer possible. In such circumstances one of the parents would remain on for a time with the unhappy youngster.

Which parent remains? When observing Black Eagle parties still assembled in Lesotho after midsummer, I noticed male juveniles with adult females and female juveniles with adult males. I therefore believed that it would be the mother's duty to remain with her son until he was settled. For the father this duty would probably be harder and his impatience could (as I have seen) lead to physical force being used to drive the young male away. But if a daughter had to be left on her own it would be father's parental duty to see that she was safe before he left to join his mate.

Observations on Eagle Hill, however, have tended to disprove this view. Ave has returned first each year (except in 1967, after an unsuccessful breeding season when both eagles returned together) regardless of the sex of the juvenile left behind. It is probable that the male returns first in order to defend his eyrie. A mature tercel would not attack a mature female alone or accompanied by a juvenile of either sex unless they entered his territory after breeding activity had begun. Another tercel entering his territory would, however, always be suspect. For these reasons the duty of seeing a juvenile settled probably falls on the mother eagle. The mature eagles which I observed with immature birds of the opposite sex in Lesotho were probably not parents but prospective mates. Should an old bird lose its mate it will soon find a new one, young or old, and bring it to its eyrie.

How do we distinguish male from female immature Black Eagles in the field? Plumage does not help. From the time the juveniles leave the nesting territory until adult life begins both sexes go through plumage changes which are not distinctive. Size is the only safe field character for sex determination of Black Eagles as seen in nature. Note particularly their wing spans when they fly together, preferably seen tip to tip. A female Black Eagle of five months old has a wingspan longer than her father's and almost as long as her mother's. A male of the same age has a slightly shorter spread than his father's.

Of the hunch that I had about Ave's return, I would mention that I have had such hunches before in my associations with wild creatures and they have seldom failed me. Could it be that there is still inherent in modern man's make-up a latent faculty for communication with other animals as they are able, and observed, to communicate between themselves? Was such a faculty well developed long ago when men lived natural lives as members of a closely-knit, interdependent animal kingdom? Is it possible to educe it again when one lives close to Nature?

These questions, arising out of inexplicable experiences, present a challenge to naturalists. In some instances I think that the explanation is simply that we study all the relevant facts and, when they point to the likelihood of an event happening, we act on the assumption that it will happen. Then we are sometimes surprised and mystified at the result of our own reasoning.

In other cases the sub-conscious mind of man does the work for him. For example, the Cape Coloured fishermen of this region are very dependent upon fish as a source of protein. When the east wind blows there is often a run of big game fish inshore off the east end of Robbe Berg. Sometimes the wind gets up in the dead of night but at first light the "Capies" will be there for the early run. Ask them how they knew the fish would be in and they will reply that they heard it in the night when asleep. But nothing will induce them to say who told them or to discuss the matter further. No doubt the voice of the wind was speaking and intuition brought the message; but only because of their daily contact with and shrewd knowledge of matters pertaining to fishing.

Then we men sometimes forget that the sensory perceptions of some animals, particularly their senses of sight, hearing and smell, are more efficient than our own. Dogs, for example, have better ears as well as noses, for they can hear sounds pitched too high for human ears. Animals and birds may therefore communicate in the normal way without our detecting their means of communication, unless we use special apparatus. Bushmen and other primitive peoples living in their natural surroundings also have very keen perception, due partly to being gifted and partly to constant use and training.

Yet it cannot be denied that animals, including birds, have a method of distant communication far beyond the range of ordinary sensory perception. This is sometimes referred to as a sixth sense by which messages from afar or warnings of impending danger are received. And the communications are two-way where mutual interests and sympathy are involved. The phenomenon of telepathic communication in man's relations with animals, as well as other human beings, is difficult to refute and experience is apparently not confined to those susceptible to psychic influence.

The pleasures of knowing Nature for man are also restricted in other ways. With eyes he strives to seek and enjoy them: through training his ears and senses of smell and touch he may enjoy them more. But with inner senses still unattuned he must forgo much.

Thoughts such as these keep me occupied during my watches on Eagle Hill but Nature always presses her claims on my attention with other offerings. During the midday watch the butterflies come flitting over the top, some to stay and sport or flirt. First arrivals are the Painted Lady and Yellow Pansy, soon to be joined by the South Coast Widow and Table Mountain Beauty, all Nymphalidae. They assemble in the jeep track clearing which is their Royal Mile for along it will soon pass the two fast-travelling African Monarchs, one a Golden Danaid, the Christmas butterfly, and the other a Swallowtail. Then come Marbled Elfs, Skippers, Blues and Coppers. The flashy Protea Scarlet (*Capys alphaeus*) has come back to Eagle Hill following the re-introduction of many species of Protea on which they breed. The eggs are laid on the Protea flowers and the larvae eat their way in and there pupate. Handsome, strong, fleet of wing and tireless these flies excel at aerobatics, butterfly style.

The Proteas have also attracted the fascinating Longtailed Sugarbirds (*Promerops cafer*), which feed and nest on them, and the beautiful Orange-breasted Sunbirds (*Anthobaphes violacea*) which feed on the nectar of both Protea and Erica and nest in the heaths and scrub on Eagle Hill. Cape

Francolin Pheasants and Redwing Partridges have come to live in the grassy clearings all round Eagle Hill; and wildfowl on the dams. The Black Eagles never touch them.

The views from Eagle Hill are magnificent, subject always to conspiracy of sun and cloud.

On the northern skyline are the sharp profiles of mountain chains for mile after mile; pyramid and cone, peak and depression, loop and crotchet, slope and line in a crinkled, folded jigsaw puzzle: three full ranges from the Outeniqua in the west, through the Langkloofberge and on to the Tzitzikama, which dip down into the sea at Storms River Mouth in the east.

In the middle distance lie the blue waters of the bay beyond Cape Seal, fringed by beach and surf as far as the eye can see, from Keurboomstrand to Nature's Valley.

Turning to the south, we face a seascape of incomparable beauty. Before us is the ocean beneath bold coastal cliffs and headlands. Many sea birds pass by, winging their way across Still Bay, sheltered by its massive cliff walls; then along the coast past coves and inlets and on to Robbe Berg Island.

The ocean scene is never the same from day to day. There are days when the pulse of the sea beats deep and slowly, when its surge leaves only a thin fringe of foam against the feet of the cliffs. And then there are days when it quickens, when the lines of waves march endlessly from the horizon and the wind is like the sound of a sword.

On cliff crags high above all this are Ave's and Mavi's two nests. And here I have studied them.

Over half a century of experience in observing breeding birds of prey, including almost all the eagles of southern and central Africa, has taught me two cardinal principles of successful observation. Firstly, the nesting birds must not be disturbed nor even be made to feel uneasy by the presence of the observer.

My eagles carry on their lives quite normally so that I observe free action and uninhibited behaviour. Forbidden are acts such as frightening the sitting bird off the nest in order to record the exact dates on which the eggs are laid or the chicks hatched. Removal of an egg or chick for weighing or ringing may be justified if the object is to establish certain facts regardless of the birds' interests and sound general observation of their family life. I spare no pains to win the complete confidence of my eagles in a natural friendship.

The second principle which I follow when studying behaviour is to interpret the facts observed after full consideration of the circumstances of their occurrence, noting the weather and any extraneous influences which might effect findings.

Observation and photography are done from hides or blinds. Hide I on the cliff edge covers the main nest and west arm of the nesting cliff, with the sea beyond. Its orientation is south. It is 170 feet from the nest and about 35 feet higher. Hide II covers the alternative nest and is about 350 feet from it and 30 feet higher. It commands a good view of the whole nesting cliff, including the main nest, and of Eagle Hill, looking west, and of the sea.

CHAPTER FOUR

My friends, the eagles

I shall always be grateful that I lived in places and times when field sports and natural history could be all-absorbing interests. As a boy I had the good fortune to live in Lesotho where we were free to roam anywhere over unfenced communal veld.

We organised ourselves as a voortrekker party, using Sir Percy Fitzpatrick's "Jock of the Bushveld" as our guide, and were encouraged to learn veld-craft by understanding parents. We had been given ponies and were taught to ride by a veterinary surgeon renowned for improving the Basuto pony as a breed. Whenever he saw daylight between the boy's bottom and trotting horse's bare back, the end of his riding crop would close the gap.

We were sent off with hard rations and told to be back by dark. We drank from mountain springs and shot speckled rock pigeon and red mountain hares for lunch with our .22 rifles. Often we would knee-halter the ponies in the sweet grass of a mountain kloof and assemble under the shade of the highest rocks on the mountain top at midday. There we would roast pigeons and rest for a while, watching the eagles and vultures soaring high overhead and training our eyes by following them until they vanished; or we searched the blue horizons for them to appear as distant specks. We learnt to recognise birds by their shape and action in flight, to distinguish between the rare fast-travelling lammergeyer and the common Cape Vulture; and between adult Black Eagles and Cape Ravens or between immature eagles and Jackal Buzzards in the far distance.

After lunch we would shoot more rock pigeon round the krantzes and lose some to Peregrine Falcons which dashed in to snatch them as they tumbled down the cliff-face. In the wheat fields on the way home we saw Lanner Falcons follow flushed quail and bind to them in fast flight. We watched Rock and Greater kestrels hover and drop down on locusts and we studied the plumage of migrant kestrels and hobbies when they came in hundreds to roost on the tall gum trees during summer.

We knew where to find nests of all the resident birds; their breeding seasons and plumage and behaviour when breeding.

At school in the rural midlands of Natal, the local farmers gave us free run of their farms in a land of unspoilt evergreen forests, hills, rolling downs and marshes. During school and later holidays I ranged further and gained experience of the birds of Natal and Zululand from the Berg to the sea and Kosi Bay in the north. Raptors, birds of prey, became my special interest at this time. I observed nests of five Crowned Eagles, sometimes spiking up forty to sixty foot tree trunks to reach the lowest branches on the climb to the nest. Long-crested Eagles (*Lophatuës occipitalis*), Secretary Birds, Jackal Buzzards, Black-shouldered Kites, Kestrels, Sparrowhawks, Goshawks and Marsh Harriers were studied at their nests.

Among many exciting experiences of that time my most vivid memories are of Crowned Eagles (*Stephanoaëtus coronatus*) hunting in the Karkloof Forests. On one occasion a troop of monkeys, which I had disturbed, climbed into the canopy of a tall tree to hide. An eagle swept over the tree, driving them down again to the open branches for safety. Then swiftly it executed a rolling loop and came back zig-zagging through the branches to pick off a bewildered monkey. I was responsible, quite unintentionally, for the poor thing's capture because the eagle took advantage of my presence to cover its attack.

In my student days at Rhodes University I worked through the country lying between the Fish and Bushman's Rivers and later the Alexandria Forest, observing specially the eagles. Penetrating the dense and often thorny bush and sleeping in caves along these rivers was hard work but very rewarding and I shall always cherish memories of days spent among the birds of this region. At first light one winter's dawn I was watching a pair of nesting Crowned Eagles start their day from my cave in a cliff above a horseshoe bend in the river when I spotted a pair of Fish Eagles at their nest in a tall tree near the water's edge. I watched these eagles happily sharing overlapping territories all day. The Fish Eagles kept strictly to the river and avoided flying over their powerful neighbour's nest. The absence of any serious competition for food between the two species may explain the lack of aggression in this instance. After sharing this beautiful river bend all day with the eagles and other forest animals, I spent another night in the cave with barn owls passing silently through the flickering light of my camp fire.

Many years later I took Sir Charles Belcher, author of "Birds of Nyasaland", to this spot by boat, and we found the Crowned Eagles with a fledgling occupying the Fish Eagle's lodge. Their own castle on its noble tree had been destroyed by a hurricane. The Fish Eagles must then have been dislodged, but they were unlikely to have gone tamely.

My survey of the Eastern Cape eagles was assisted by interested and hospitable farmers and foresters with whom I stayed. Dr. John Hewitt, then Director of the Albany

Museum in Grahamstown, always encouraged field studies and Dr. Leonard Gill, then Director of the South African Museum in Cape Town, once gave me a lesson in identification of waders on the Cape Flats.

My introduction to the birds of Rhodesia and Zambia was through a university cricket tour. Afterwards I stayed on to investigate Bateleurs at Broken Hill and Black Eagles in the Matopos, where some friends took me camping by mule cart. After six more years in Natal and Lesotho and two trips through Zululand with Kenneth Pennington of butterfly fame, I went north again in 1930 and stayed for thirty-five years.

In Zambia forty years ago wild animals and birds still lived relatively undisturbed in sparsely populated country where they had freedom of movement. African hunters who followed the spoor killed only a few with their muzzle loaders. The Kafue Flats and surrounding country, the Lukanga Swamps, between Broken Hill and the present Kafue National Park, and parts of Barotseland, notably the western watershed plains and the upper Zambezi basin flood plains and tributaries, were game areas which I knew well. During the summer rains the game scattered and moved about through the higher bush country.

Around these localities of natural game concentration the birds of prey lived and bred in great variety and numbers. Waterbirds and wildfowl from all over Africa and beyond also flocked to the inundated plains in the rainy and flood seasons. Some species had their permanent homes there, others came to breed and many were seasonal visitors.

Here Fish Eagles (*Haliaëtus vocifer*) were seldom out of sight or hearing. Perched or overhead they displayed their rich chestnut waist sashes on pure white bodies and epaulettes of the same colour on black wings or proclaimed their dominion with wild yodelling cries. They fished the shallower waters, often ending their stoops with a plunge; then, clutching the fish with the claws of one foot, carried them away head first to their large nests in the upper main forks of big trees. This was their land of plenty, of bream and barbel, Zambezi tiger fish and Kafue pike. Well fed, they produced larger eggs than the South African birds and clutches of three, instead of the normal two, were not uncommon. The long laying season covered four months, from May to August.

Tawny Eagles also bred freely in the huge Acacia trees along the fringes of the plains and in the surrounding savannah. These quite large rufous brown eagles are of pure Aquila stock and closely related to the Golden and Black eagles though rather smaller and perhaps less regal in bearing. Living in hot and often dry bushveld they can be sluggish and ignoble in their feeding habits. They will scavenge pulpy remains of mammals, birds and reptiles killed on the roads by speeding cars and sometimes hang around carrion carcasses in the company of vultures and marabou storks. Bateleurs also feed on carrion at times.

I once kept an adult Tawny Eagle (*Aquila rapax*) which was found with a wing wound on the edge of the Kafue Flats. Tor soon became friendly and amazingly tractable. She loved riding on the spare wheel at the side of my two-seater Dodge and I took her out of an evening to shoot hares along tracks in the grassveld. When a hare was shot she was taken to pick it up; then brought back to her tyred perch and driven home still clutching her prize. She soon learnt to dash forward on her own and pounce on the kicking hares. When completely recovered from her wound she took off from the running car to capture startled prey.

Far from being a malingerer, Tor soon won her spurs in my service and proved her blue blood. She adapted herself in adversity to a new way of life with remarkable application.

She was completely free but remained with me for a time, coming and going from the thorn trees in my garden. She will never know of the happiness she gave by her simple trust.

Near Mazabuka I collected a beautifully marked clutch of two Tawny Eagles' eggs at considerable risk. A swarm of wild bees had made their nest under the eagles', attaching the combs to the closely packed sticks. The day was hot and the bees very active so I waited for the cooler evening air. The nest was in a high fork of a sloping lateral branch in a huge thorn tree. I reached it safely, packed the eggs in my leather collecting box and slung it on my back but as I began the descent the bees suddenly attacked. I slithered down the rough-barked tree like greased lightning, losing strips of skin from insides of knees and wrists and receiving many stings. Once on the ground I escaped the angry bees by retreating into a dark thicket. There my African assistant quickly scraped off the stings with a sharp knife, avoiding compression of the poison sacs. I had kept bees as a boy and knew the drill.

This set of eggs was richly marked with shades of lilac and mauve instead of the usual brown or rust coloured markings. I have taken eggs of Martial and Wahlberg's eagles and of the larger African vultures with the same rare and attractive markings. Three weeks later these Tawny Eagles had produced another clutch which they successfully reared, despite the bees in their basement.

I have also known Black Eagles to live happily in close proximity to bees. Their nest was built on a krantz ledge above a deep vertical cleft containing an enormous bees' nest. Bushmen had driven hardwood stakes into the cleft to rob the bees and there they had remained for a great many years.

Weavers (*Ploceinae*) which breed in colonies will sometimes suspend their nests from trees on which there are eagles' nests. This may be deliberate in order to obtain the protection of the eagles from enemies such as snakes, monkeys and harrier-hawks: I have noticed such associations with Tawny, Fish and Wahlberg's eagles and Bateleurs.

Whalberg's Eagle (*Aquila wahlbergi*), a small and lively dark brown eagle, came to breed in Zambia, mainly in September and October, in very large numbers together with the African Black Kite (*Milvus migrans parasitus*). The former laid their single eggs, which vary greatly in markings and size, in the wooded hilly country whilst the fork-tailed, yellow-billed kites nested everywhere, laying three well-marked eggs in nests always lined with the smelly faeces of jackals and other carnivorous animals.

A pair of Booted Eagles (*Hieraaëtus pennatus*) was found nesting at Mongu in Barotseland. This eagle is a summer migrant from eastern Europe and western Asia where it breeds. Breeding in Central Africa is extremely rare and such records should be of great interest to scientists as well as field naturalists.

The common White-backed Vultures (*Gyps africanus*) nested on trees in scattered colonies and the Black or Lappet-faced Vultures (*Torgos tracheliotus*) and White-headed Vultures (*Trigonoceps occipitalis*) away on their own as a rule. The latter is eagle-like in flight and will kill smaller game and birds after the manner of an eagle as well as feed on carrion in the company of other vultures. The Martial Eagle, three species of Hawk-Eagle and three of Harrier-Eagle, the Gymnogene (*Polyboroides typus typus*), the Secretary Bird, African Marsh Harrier and many smaller raptorial birds were observed nesting in localities which suited their requirements and were generally more common where game was plentiful.

Of all the Zambian birds of prey, the Bateleurs (*Terathopius ecaudatus*) were the commonest and most generally distributed permanent residents. Found in all countries south of the Sahara, the Bateleur is very well known in Africa and one of its most strikingly characteristic birds. In boldness of form and colour this is an uncommonly fine and robust bird of a very different type to the Secretary Bird which is certainly its strongest raptor rival for African heraldic honours.

I see the Secretary Bird as a tall slender and very distinguished looking marching eagle in grey and black uniform with court plume. He is the guardsman of the African grasslands.

The Bateleur may, by the same token, be regarded as the patrolman of the wooded country but so sprightly is he that one could also portray him as the scarlet pimpernel of our eagles, in glossy black cloak and red cape with grey wing shoulders and crested cockade.

Patrolling his wide beat high overhead in fast straight flight by day, he keeps a sharp look-out for any creature, living or dead, in the bush below. When searching for prey he examines the cover in detail by making use of a natural tendency to rock or roll from side to side on his exceptionally long and broad upswept wings, so penetrating the field of vision from many angles. Another hunting device is to cast his eyes backwards over the field already covered, viewing it from underneath his body, with bowed head and arched neck, in order to spot animals that give their position away by moving for cover after they have seen him pass over. Snakes and slow moving quarry are tackled with caution, after a sidewards drop, but small fast mammals are taken in a grand slam stoop made quite devastating by its sibilant whine.

Le Vaillant named this bird Le Bateleur (Harlequin or Mountebank) probably because of its way of turning somersaults and performing other aerobatics during courtship displays. The act is funny for some people, delightful to others.

But Bateleurs are not professional buffoons. I found them extremely shy at nesting time. The male approaches the nest secretively and guards it well from the air or vantage points, warning the female of danger by barking an explosive "caw". She will then leave the nest unobserved and slip away between the trees or down a glade. A large tree on the fringe of a glade is favoured for the rather small but deep nest which is placed either in a main upper fork or on a stout prong high in a lateral branch. One large creamy white egg is laid early in the year in Zambia, usually during February to April. Bateleurs differ, in my experience, from many raptors in not occupying territories from which other birds of prey are excluded. Young Bateleurs, still in brown plumage, are often to be seen in the vicinity of a nest in use. Later in the year Wahlberg's and other eagles, Kites and Lizard Buzzards will be allowed to nest in the area. A pair of Bateleurs I knew permitted Lanner Falcons (*Falco biarmicus biarmicus*) to use their nest when they had finished breeding. My conclusions are that Bateleurs do not claim territory for their exclusive use but rather that they have hunting beats recognised by other pairs of adult Bateleurs.

Although Bateleurs disappeared from the Western Cape many years ago I have seen three individuals in the Knysna district in recent years. Strays would be their official listing but my view was that they were scouts. And who would not welcome Bateleur patrols in our rodent-infested areas.

Eagles have been my great interest, my passion even, for many years. And the most magnificent of them all is the Black Eagle.

I know the Black Eagle as the dashing beau sabreur of kransland. I study him sitting thoughtfully in his academic robes and know him for an ancient philosopher. I examine him in flight mastery and in life's strife for perfection and laud him to the skies. I crown him Eagle Lord of the African Skies and hail him: *Sic Itur Ad Astra.*

CHAPTER FIVE

Eagle courtship

After Ave had been home for ten days he disappeared and returned two days later with Mavi. Once again they must, I could see, have taken the Formosa Peak route. They came romping along together, their two double-wing-shaped forms sometimes merging in silhouette as they rode buoyantly on the air stream. And surely their spirits were uplifted in the annually recurrent elixir of a new mating season. Absorbed only in one another they alighted on their favourite rock perch overlooking the eyrie and, sitting side by side, stroked beaks before relaxing in the cool sea air. These eagle masters of mobility must rest often to gain the strength required for their very high rate of metabolism.

For a few weeks only will the eagles enjoy respite from the labours of procreation. Below them when they sleep on their night perches above their nests are ocean waves that never rest. Yielding to the pull of moon and sun, they symbolise the endless rhythm of changing seasons. Now summer will soon be moving at the sun's behest to northern lands. Throughout the autumn, winter and spring the eagles will be drawn, as inevitably, into a series of activities designed to ensure not not only survival but regeneration of their species. So the eagles will play their part in the living world's oldest natural process of selective evolution.

At this early stage the eagles are seldom home, for they do their hunting in the remoter parts of their territory and even beyond its borders, as most other birds of prey are not yet defending their territories. The Formosa Mountain area in the Tzitzikama is a favourite haunt on hot days.

So, for a time, the dassies which live near the eyrie may safely sun themselves on their rock ledges.

The Rock Dassie (*Procavia capensis:* Family *Hyracoidea*) also called Hyrax or Cony or Rock-Rabbit, is the principal and staple food of the Black Eagles today.

This mammal normally lives on krantzes or cliffs and in rocks and scree on hills or along water courses. But so rapidly are they increasing in some areas that they overflow their natural habitat and spread into earth holes or thickets on the flats. Families of four to six are born each spring when Black Eagles are feeding their fledglings. The weight of adults is nine pounds. They feed on much the same vegetation as sheep and four dassies are capable of consuming as much as one sheep. They are tough and hard to control and very destructive.

When the eagles are away hunting dassies far afield they are sometimes absent for two or three days. On other days they leave their night perches above the eyrie early and return in the twilight well after sunset.

But they evidently keep a watch on their territory, because unwelcome raptors do not invade it when they are absent, except perhaps for a vagrant peregrine falcon passing along the cliffs and the ubiquitous kestrels. Ave comes back to the eyrie more frequently than Mavi, probably to check up.

Eagle Hill eyrie

The south side of Eagle Hill is sheer cliff, most of it 300

feet high and rising straight out of deep sea, except for the eyrie crescent itself which is set back into the hill. Cut and remove a very thick slice from a round cake and you have a model. But the slice has not been cleanly cut for the full depth. Below the face of the north wing cliff and the upper part of the west wing is a broad, land shelf sloping fairly steeply down into the sea. On it are rocks and scree and minor cliffs, interspersed with terraces covered with low succulent vegetation. This secluded place below the cliffs will be Eaglet's home, school and playground for nearly two months after he has left the nest. Looking east from the nesting cliffs is a grand view of the coastline and to the south is the open sea with ships passing by, day and night.

Many sea birds live on the rocks below the eyrie or visit the eagles' home waters. Black oystercatchers on sturdy red legs run up and down with the waves, probing rock fissures for shell-fish with sharp bright orange beaks or cleaning-up sea urchins on shelly beaches. White-breasted cormorants dive in the tidal pools or sit about on high rocks. Fourteen pairs of them nest on narrow, white-washed ledges below the eagles' main nest and fly to and fro all day feeding greedy young by regurgitation. Cape gannets, following shoals along the coast, weave up and down twenty feet above the waves; then sweep up high and roll round for a deep plunge to poach a mackerel, mullet or pilchard under water.

Far out to sea, emerging white as it moves on the horizon, appears a single bird, then another. The avian world's most marvellous mariners and Neptune's favoured fliers, renowned for their epic odysseys, these graceful, slender-winged albatrosses, spreading eleven foot of sail from tip to tip, dip down troughs and rise over crests. Lonely, happy wanderers of the vast southern oceans, they come to caress the waves in the wake of ships. From their eyrie the eagles watch.

Nest repair

At midday on the 12th March both the eagles alighted on the main nest and began clearing the central saucer where the incubation basin, of about eighteen inches in diameter and ten in depth, would later be specially prepared for the eggs. They use their beaks for moving and placing material in position on the nest but will occasionally drag a large stick, which has caught up, with their feet. All material is carried to the nest in their feet.

The main nest is a massive structure, built up over many years, of dry sticks, intertwined in a tangled mass, on a ledge under an overhanging rock. The ledge slopes upwards in the direction of the sea and the nest is about five feet high from base to top. When the nest is in repair the top platform is three feet across, including a section under the overhang, which is not visible, and the landing stage on the open east side.

Along the southern perimeter is a projecting rock forming a dwarf wall which protects the platform from the south wind. On the west side is the cliff wall sheltering the nest from the prevalent south-west wind and rain. The nest is situated roughly midway along the west wing of the eyrie cliff. It is thirty feet below the cliff-rim and approximately 240 feet above the cliff's base. The south end of this cliff stands in the deep sea where the waves have carved out a cavern resembling the arch of a human foot. Here the 300 foot high cliff bears the name of South Point. Round this point a sheer cliff wall rises from the sea and runs westward until the line is broken by the Still Bay basin, 400 yards distant.

During the next ten weeks the eagles slowly repaired the main structure of the nest, selecting dry sticks, some up to three feet long, and brush in the fynbos scrub on Eagle Hill and carrying these, one at a time, in one of their feet to the nest. Many sticks are placed round the platform to make a solid rim on the exposed east side and the sloping north side, which is protected from north winds by projecting cliff some forty yards away. On this prominence are the family's separate night perches. These face south-east and overlook the main nest, coast and sea.

For the first six weeks, nest repair proceeded slowly. In some weeks two or three sticks were added, in others none at all.

From the 21st April the pace was increased and the eagles sometimes worked as a team. Both would sit together on their courting perch at the east end of the north wing of the nesting cliff until the morning sun warmed their world and provided them with blessed air currents. Ave would then show off his skill in aerial display above the perch and after he had returned to it and perhaps stroked beaks with Mavi, both would take off and frolic about in the air. Then each of them would drop down separately to collect a stick and carry it to the nest. Now Mavi remained on the nest and fussed about the placing of the sticks whilst Ave fetched more. She picked up sticks which he had brought and carefully placed, tried them in other positions and usually ended up replacing them in Ave's original position. Ave is very good natured about all this concern. Understanding and affection, though not overtly demonstrative at this stage, are clearly discernible even to human beings who do not know Black Eagles. These happy home builders have a beautiful companionship.

Fine, springy sprays of heath, some still bearing dead leaves, are used for the basin. Mavi usually does all the work of fetching and arranging these while Ave is away hunting. But later on it is she who is often away hunting and feeding well in preparation for egg production. Then there are days when Ave alone works on nest repair jobs.

Lining the nest

On the 20th May, or perhaps a day earlier, the eagles began lining the central incubation basin with freshly plucked, green leafy twigs. These are pressed down by Mavi with her feet to form a mat which will cover the floor and sides of the basin. The leaves will supply the moisture necessary to produce humidity needed for successful incubation of the eggs. The leafy twigs will also provide a soft, smooth, wind-proof carpet covering the rough dry sticks on the floor of the nest and will prevent scraps of food from dropping out of reach between the sticks.

Black Eagles are meticulous about their nest hygiene. No food remains whatsoever are left about to putrefy and attract flies, ants or other insects. Twigs which become blood-stained during a meal on the nest are removed and replaced by fresh ones. A regular supply of green sprigs is maintained throughout the incubation period in order to ensure sufficient moisture for the eggs. Thereafter twigs are brought from time to time as required. The last spray was brought on the 9th October when Eaglet was 83 days old and almost ready to leave the nest. The old green twigs fade and the brown and yellow quilt-work patterns made by the leaves provide a very effective camouflage for older nestlings. Green materials are picked from Rooikrans (*Acacia cyclops*) or pine trees (*Pinus*—various species). The eagles drop down and alight on upper branches, pluck a twig with their beaks and either carry it to the nest in their beaks or occasionally, in heavy winds, transfer it to claws when flying off with it to the nest. In light winds they hover over the tree, grab a spray with a foot and fly on. Small branches of rhinosterbos and heaths are plucked when standing on the ground or rocks. The latter are usually placed on the platform just outside the incubation basin.

I know that most diurnal birds of prey (excluding owls)

line their nests with green twigs. But I have often wondered why the Black Eagles use alien pine and Rooikrans for green material when suitable indigenous trees are readily available. Are they preferred because they keep longer or is there some other reason?

Following is a typical entry in my records which covers fifteen minutes of observation on the 20th May, 1968: "The female remained on the nest all the time . . . The male came to the nest twice with green twigs and placed them carefully whilst the female merely moved out of the way and watched the male. As soon as the male left she walked about in the basin, trampling down the lining materials and occasionally adjusting a twig which was not placed to her satisfaction. She was very fussy."

On the 29th and 30th May Mavi fetched some dry water grass, sedge and rush leaves from a dam half a mile to the east of the eyrie. The long yellow leaves floated behind her in the breeze like streamers when she carried them to the nest. She used them in the basin for a final layer of soft padding. Meanwhile Ave was busy hunting and brought back a dassie for her each day.

It is interesting to note the division of labour for nest building. Mavi was responsible for collecting and arranging many of the dry twigs used for the first lining of the incubation basin and all the soft materials for the final padding. Ave had carried most, but not all, of the heavy sticks and brush to the nest and they shared the work of collecting green twigs for the basin. The use of waterside vegetation for the basin probably dates back to primeval times.

Should a Black Eagle of either sex lose its mate, it will repair its nest alone when the time for breeding comes round. Meanwhile it will be on the lookout for a mate to bring to its eyrie. I have known a Black Eagle male repair his nest to the stage of bringing the first green twigs before he arrived back one afternoon with a new mate.

On the last day of May Mavi became broody.

Courtship and mating

Whilst the basin is being lined the eagles' courtship play increases in frequency and intensity and mating takes place on several (perhaps ten or more) occasions, sometimes on their favourite rock perch at the east end of the eyrie cliff or South Point, but not always. On one day it was observed to occur twice, once on each of the points mentioned. They begin by sitting quietly together on one of these perches in the warm sun. Then Ave takes wing while Mavi watches him. And so begins a Black Eagle courtship festival during which they put on thrilling flying displays and aerobatics and dancing shows. The daring stunts of the dashing male arouse the female who soon joins in the rough and tumble and dances with equal enthusiasm.

The air is more than the breath of life for Black Eagles: it is their very existence. The skies above mountain and valley and cliff-top are their love-play heavens where they harness the winds of the wild to ride the skies for their hearts' delight. Forgotten on these wedding days are work-horse winds. Present time and present mood are for fleet motion, dare-devil dash and light-fantastic wing-tread measure.

First they rollick and roll rapidly round and round, then mount a spiral high into the sky in golden silence; now they race highflying at speed, winged arrows with feathers tuned to a silken whisper; now they leap recklessly into headlong swoop and plunge like plummets with devil-dare abandon.

Gallant beau sabreur they hurtle down the high-sky-to-deep-sea incline, cliff wall to lee of them, ocean to left, breakers ahead. Yet never a flinch nor swerve. Down, down undaunted they dive gathering momentum audacious through cloven shadows with wings new burnished steel grey.

On, on they speed straight for the pounding breakers. Now at full speed in the final spurt, pinion sabres flash bare, raised high and clear as they brace air. Straight on through cliff-end barrier they crash and smack into the sou'wester's wind-wall beyond, to rocket up half a league into the sky with wings bow-sprung to absorb and then unleash, in upward thrust, the impact's rebound forces.

The exhilarated eagles roll leisurely over and, leaning on a slant of wind, first side-slip and then flutter down just like the autumn leaves falling even now in the wildflower garden for which they immediately set course, down wind and schooner rigged.

Before we watch their country dances in the valley skies, we might review their devil dive act and verify that they were really and truly playing with death. The report on this performance, as I preferred to tell it, is not proof, nor has a full account of this dangerous sport been given before. We might first look at the speedways used by the eagles for their terrific swoops. The longest which I have seen them use is the Still Bay West one, generally, but not always, favoured in a west wind or sou'wester. The shortest is along the eyrie west wing cliff line to South Point, also used in these winds. The third is the Still Bay East course which lies immediately west of Eagle Hill and is preferred in south-easters and east winds.

A description of the first will suffice. It is over a mile in length and the eagles begin at racing speed, high over Rondebosch Rivier Kloof and dive down on an incline of about 45 degrees. In the first half mile they are making pace against wind by exploiting its down thrust and gravity's pull. They swoop on, hugging the massive cliff complex, some 600 feet high and half a mile long, accelerating in its shelter, to encounter the boisterous wind violently again at sea level beyond the bluff. Apart from the scorching speed and hazards of the rough contour, many cormorants nest on these cliffs and they and numerous gulls fly across the speed track or sit on high rocks along the route. All these marine regiments and their raven camp-followers panic and scatter haphazardly before the eagles' sweeping enfilade, so endangering themselves and the eagles. Worst of all, at the crucial moment before the eagles' rocket thrust skywards when they have crossed the divide, between cliff-end and ocean, and are literally between wind and water, they are as good as blindfolded because they cannot see what is approaching round the corner and along the cliffs from the west. Always there are terrible risks of smashing into a gorged gull, cormorant, gannet, mollymauk or other sea bird beating its heavy way down wind to round the promontory. The impact of a head-on collision in mid-air over the breakers of a jet-speeding eagle and an inert lump of flesh and fish would blow them both to smithereens in a puff of feathers and fish bones.

Finally there is positive proof that the eagles themselves realise the dangers of their devil dives. Mavi refuses to participate when her time draws near and she is carrying developed eggs within her.

Also, this daring feat is the final test taken by young Black Eagles when they come of age and before they win their wings at the end of flying training. To my knowledge, Eaglet was taken three times by his parents for this test and funked it twice before he took the plunge. Sixtyfive took it after jibbing once. Sixtyseven was not ready for it and, as far as I know, was never put to the test before she left home.

To a human observer the feat is astounding. It is one measure of the courageous life of Black Eagles.

Courtship dances

My eagles go inland for their courtship play which includes the special Black Eagle dance and aerobatics. Their play-place is the sky above a valley and generally one in the mountains, for Black Eagles are naturally montane dwellers. Their displays are truly rural native country dances with aerial pageantry thrown in for good measure. Because my eagles go away for their dancing festival I am unable to attend their performance and must transport myself forty years back in memory to my native land, Lesotho, where other Black Eagles used to entertain me. Happily I am also able to rely on descriptions of Rhodesian Black Eagles' displays given to me by my friend Bill Miller.

The dance begins with the partners whirling round in tight spiral turns, the male above. Each time they start a new volution, and rise higher, the dancers clap their wings simultaneously, sometimes in single wafts or two quick beats, or three or four, together. Now the female holds her position, treading on air with gentle wing-beats, while the male rings higher: then with closed wings and locked claws he drops straight down upon her and springs back to his original position after narrowly avoiding a bump. After he has repeated this action several times, the partners change places and go through the whole movement again. The dance continues until each of the partners has had three or four turns in the active role above. The action is like that of a child's yo-yo toy which is thrown down only to rebound to his hand on its resile chord; or like a tennis ball being bounced between racquet and ground. The partners appear to be linked by an invisible elastic band as they separate and come together again.

Between dances the female dashes away fugitive, sweeping smoothly and lightly along until the male overtakes her. Then they return to the dancing place and dance again.

Black Eagles delight in many less formal and more frolicsome dances. They leap high after dives and iump over each other in saltations, with sudden transition of movement into salients or skipping in saltarellos. They chase one another round imaginary quarter-moon crescents and loop full loops. I have never seen them turning tail over head somersaults; they always roll over at the top of the arc with a neat turn of the wings and tail even when the intense excitement of an acrobatic dance has reached crescendo.

All the four winds are their playfellows when not needed as servants. Occasionally a raven will take the floor and the dancers will suffer him and his insolence for he is so obviously a ninny.

Black Eagles dance with serious concentration and never utter a sound when dancing though they take transparent delight in this pastime. To human observers there can be few more spectacular sights.

CHAPTER SIX

Mother Mavi

On windy days, and there are many, the exuberant eagles revel in the reckless ardour of their mating flights. In sportive mood, they drop off the nest together and glide down fast under the lee of the cliff, making for South Point at midway height, then into the teeth of the west wind along the sea cliffs to Still Bay. Now they drive themselves forward furiously, intent on plundering the wind's power by plunging their lanceolate bodies like arrows into it and rising up against it for another plunge and another and yet another and another. After each successive dive down to sea level, and bounce off a wave's springboard, their impetus carries them higher into the air above the cliff-top's rim. And so they ramp through a row of deep vertical V's, making distance and gaining height rapidly. Great skill and daring are needed for attacking the heavy winds and it is a grand sight to see.

Many times have I watched their battles against gales from my hide and admired their perfect timing and judgment and graceful action. How sadly wrong it is to claim that such flying ability is merely instinctive! I have seen all the three young birds bred here stall at the top of their V's when doing this exercise at quite advanced stages of their training. Yes, nature there must be, and of noble quality, nurtured with the help of patient Black Eagle parents and perfected by hard experience for of this, and no less, must these glorious fliers be made.

Mating

These exciting courtship flights serve a dual purpose. The eagles may have set out to collect material for their nest or to hunt at their destination. They set out in fellowship as two individuals with a life-long association, because Black Eagles mate for life. Year after year their lives are lived in an annual rhythm, the swing of which is instinct controlled. Their homebuilding has been instinctive.

And now also instinct swings them on to intercourse: the mating act follows as surely as day follows night, unless disturbance intervenes.

And so at the end of a mating flight, before the job in hand begins, they alight on a high, safe rock and sit quietly together for a few minutes perhaps. Then the male strokes beaks with the female who crouches when he mounts for coition. The act takes five to seven seconds from mounting to dismounting. Bodies are then shaken with feathers ruffled up and preening generally follows. The pair remain together for a time and the male is always the first to leave the perch. This act never takes place on the nest, just as the eagles never take food to the nest or eat it there until they have a nestling to feed.

Mating also takes place, usually on a favourite perch on the eyrie cliff, after the displays of country dancing or aerobatics which, at this season, are direct excitant preliminaries to the act and also serve the important purpose of exercising the body when the eagles are not busy on other jobs. This exercise is necessary. Very shortly before the first egg is laid, and in the interval between the laying of the first and second eggs, mating sometimes occurs without preliminary flights by the female and after only a cursory flip round by the male after the pair has sat together for a few minutes.

The eagles' flight

Although the eagles take more risks when flying at mating time, because they are showing off in reckless display and high spirits, they often make journeys against strong winds flying the Multi-V way in the manner described. Cliff-side routes are usually taken when flying in this style. As a point of great interest, I have observed that the "bounce" they obtain off air cushioned by the water's surface below sea cliffs carries them higher than one over land at the foot of a precipice or krantz. Possible explanations are that water has a better surface than earth for containing pressure exerted by the eagle at the moment of completing its swoop and bounding up from water level: or that a rising wave forces air up simultaneously with the eagle's down-push above it, thereby duplicating air compression. On the other hand the land below a precipice is covered with vegetation or rocks and the stoop is checked earlier.

The eagles have other ways of flying against strong winds. One is to fly along the cliff contour more or less on a horizontal course but rising and dipping slightly above and below the rim and taking advantage of gaps in the cliff line to rise higher on the wind streams flowing through them.

Another way is to ring high and then soar to their destination. This was described when we followed Ave on his long flight back to the eyrie.

Finally, there is the laborious method, seldom employed by adult Black Eagles, of steady wing beating in an action which may be described as rowing. Young birds, lacking flying experience, may be seen plodding along in this way.

When flying down wind, the eagles simply maintain buoyancy and desired height by the tilt and spread of their wings and tail. Now they sail on the air, using their banks of tectrices, covering the bases of flight and tail feathers, as sails. The angle and spread of the broad, square tail are important in all flying and the feet, projecting beyond the base of the tail feathers, are normally kept hard up against the tail and work with it, so strengthening it. Black Eagles are as graceful when whistling down the wind with abandon as when flying disciplined against an adverse wind.

Eggs and incubation

In 1968 Mavi laid her clutch of two eggs in peace and calm. Observation immediately before and during the first ten days of incubation had been reduced to the minimum and when possible from a distant point so that she would be at ease. Judging from her movements I estimated that the first egg was laid early on the 1st June and the second early on the 4th June.

On the 9th June, Mavi allowed Ave to relieve her on the nest for a long spell of nearly two hours in the early afternoon and next day her tension had gone and she was relaxed and care-free. At 3.40 p.m. on the 10th June she allowed me to inspect the eggs thoroughly. My record reads:

"I slipped into the hide and opened the aperture very quietly. Mavi was sitting on the nest facing eastwards. She glanced at me and a little later stood up, walked round the nest, adjusted some rushes lining the basin of the nest and rolled the eggs over. I saw both eggs very clearly. One is rather larger than the average and lightly marked all over with rust-coloured markings. The other is about average size and creamy white in colour. Mavi looked at me as if to say, 'Seen enough? Good looking aren't they?' Then she settled and soon began preening her wing feathers as she incubated the eggs very contentedly. She really was showing off the eggs to me and her acting was superb—the proud mother!"

On the 17th June at midday I again had the opportunity of seeing both eggs very clearly when Mavi glided down to the nest and took over brooding from Ave. Incubation proceeded normally and without any disturbance but on the 15th July, there was only one egg, the smaller unmarked one, in the nest, which was filmed. The larger, marked egg had been removed, almost certainly by Mavi, because I have reason to believe that no gull, raven or human nest robber was responsible. Baboons and reptiles cannot reach the nest.

Certain broad principles have emerged from my long study of the oology and relevant behaviour of diurnal birds of prey.

During the ten days or so before the eggs are laid the male is very alert and nervous about intruders. If the nuisance continues the nest may be deserted. During the first ten days or so of the incubation period the female is extremely sensitive to the danger of potential predators and nests and eggs may also be deserted at this stage. Once incubation is well started, however, raptors will carry on breeding with determination, despite much inconvenience and disturbance. But obviously if the eggs get too cold the embryos will die. This principle may not be fully applicable for eagles that have grown accustomed to disturbances. Black Eagles are very adaptable.

During the early stages of her productive cycle the female is able to produce a second clutch, and even a third, within two to six weeks of losing or deserting the first clutch, provided that the first clutch had not been incubated to an advanced stage, somewhere beyond half-set. The second or third clutch may be laid in the original nest, if the nuisance has abated meanwhile, or in a recognised alternate nest, or a previously abandoned nest, or even a new one, hastily constructed. If eggs must be collected, under permit for scientific purposes, they should therefore be taken fresh, and with minimum disturbance, so that the female may soon lay again and little harm will result. But if insufficient time remains of the breeding season for adequate education of the young, they will suffer.

Not only are the eggs of each species distinctive, despite considerable range of variation in markings (species which lay unmarked eggs excluded), size and shape but each individual female produces clutches peculiarly her own in pattern. An experienced oologist is able to distinguish individual mothers by their clutch patterns, despite some fading of marking on eggs laid by very old birds. This knowledge is useful for tracing local movements when nests are deserted or migrations. The eggs of all diurnal birds of prey have a green-tinged inside lining when blown and viewed through the hole against the light, whether the "tissue" or egg membrane has been removed or not. Owls' eggs have yellow linings.

For species which normally lay two eggs, such as the Black Eagle, the first-laid egg is larger and better marked than the second and produces a female chick, whereas the second-laid egg produces a male. In the case of the Black Eagle, the male chick hatches three days later than the female and is therefore appreciably smaller and much weaker than she is when he hatches, because a chick grows much in three days.

Two eggs but one eaglet

Now we may pose the question: Why did Mavi remove the larger egg which would have produced a female eaglet? In the absence of the disturbance factor we may assume that it was fertile. The first answer is that she removed one egg because a Black Eagle pair is quite unable to educate more than one eaglet at a time. As the story of Eaglet's life will show, they are fully occupied with bringing up and training one eaglet during the nestling and post-nestling period at home, totalling 177 days, and a further period of over a month while they see it settled outside their home territory.

It is abundantly clear that the mother and father could not cope with two eaglets, each demanding all their attention, if they are to be adequately trained for Black Eagle life.

I realise that this is a statement of major importance because ornithologists have hitherto stressed the economic factor, namely that the eagles are unable to provide or feed enough suitable food to two eaglets at once. The provision of enough food is no problem at all for most Black Eagle families, unless man or disease has killed off all the dassies in a particular area. The feeding of two nestlings until they are able to eat by themselves would mean more time on the nest for the mother, who invariably eats her own food on the nest both when feeding young and after the young one is replete. It is not an insuperable problem. When the young are starved there is always a good reason. After the not-fully-grown eaglet has left the nest and is living on the terrace below it, he or she has to be taught to fly properly, to kill and hunt and to take good care of himself or herself in a way of life that is dangerous. It is manifest that the parents could not train properly two wilful youngsters each at slightly different stages of development and wanting to go separate ways. The Black Eagle way of life is highly complex and they are products of a very advanced stage of animal evolution. In the story of Eaglet's life I shall endeavour to show beyond all question that the educational commitment is the principal reason for Black Eagles raising only one eaglet.

Why, then are two eggs laid? This is a provision of Nature to insure against accident. Nature is very free with eggs. It is a cheap way of doubling the chances of one hatching safely, just as a very cautious man puts on a belt and braces. For the same reason both eggs can be replaced at short notice.

Did Mavi knowingly choose to have a male eaglet? I

believe she did. Why? It seems reasonably certain that she was aware that if both eggs hatched the female chick would survive. That had actually happened the year before. Sixty-seven was one of two siblings born in the alternative nest. They were seen for the first time by me on the morning of the 23rd August 1967 and as Mavi had continued incubating until then it is probable that the second chick hatched early that morning. I saw her feeding both chicks on the 24th, 28th, 29th and 30th August. On the 31st August filming of them began and the photographers reported that there was only one chick in the nest.

What happened to the other chick is not known but, according to most observers, one chick kills the other and the parents condone its action. During her life Mavi, whom I estimate to be about 50 years old and certainly older than Ave, must have seen this happen to her chicks many times and would be aware that it was the female chick which survived when she allowed both eggs to hatch. The reasons for the female nestling surviving are that it is born three days earlier and is stronger, bigger and heavier and therefore able to win the battle for food and life in which the chicks instinctively engage.

Sixtyfive was first seen on the main nest alone when he was discovered on the 2nd July, 1965. He was then estimated to be a few days old. It is very probable that Mavi had removed the other egg before it hatched.

We know for certain that Sixtyfive and Eaglet were tercels and Sixtyseven a female when they left the nesting area as ring-tails. I was extremely fortunate in having been able to observe each of them in turn flying overhead in line abreast and tip to tip between their parents and there was no question whatsoever that Sixtyseven's wingspan was appreciably longer than Ave's, whereas the others' spreads were slightly shorter. There was no possibility of my confusing them with other eagle family parties. Sixtyseven also displayed other field distinguishing marks which I have noted for confirmation when I am able to observe another female juvenile in future. More observation is necessary before it can be stated emphatically that when two Black Eaglets hatch as siblings the survivor is generally a female, but I believe it will prove to be so after a fuller investigation.

Another question of great interest arises. Is Mavi purposefully regulating her output of male and female eaglets? And if so, is she aware of the disproportion of the sexes in the younger generation of her species in the region known to her? These questions touch on what may prove to be a valuable new field of study for naturalists.

The statement that the first-laid egg is the bigger and better marked and produces a female needs further qualification. It is limited to my own experience of raptors which normally lay clutches of two and by the fact that some of them lay unmarked eggs and that species, such as the Black Eagle, which normally lay marked eggs, may lay sets containing one or even two unmarked eggs. Also a female laying for the first time may lay only one small egg, usually well-marked, and a very old bird one large egg. When a nest is found to contain one egg, another egg may already have been removed by the eagle or been lost accidentally.

Any study of correlation of egg-type with chick's sex should include raptors which normally lay clutches of one, such as Wahlberg's eagle, the martial, the bateleur and species of vulture.

Incubation duties

The female Black Eagle is mainly responsible for incubating the eggs and the extent to which the male relieves her varies with individual pairs and other factors, including disturbance and weather. Mavi does all the brooding at night whilst Ave sleeps on his night rock-perch north of the nest. Mavi usually remains on the nest until it is in shadow at 10.00 a.m., with a swing of approximately one hour as the days grow shorter or longer during the incubation period. She looks up anxiously past the rock projection awaiting the welcome shade every morning. Then she is ready to leave the nest provided Ave is back with a kill and can take over incubation.

For the first eight days of laying and incubation and the last two or three days before the egg, or the first one, hatches and until the second is hatched, if there is one, she is reluctant to leave the nest or be away for long.

At these times Ave brings her food to the dwarf rock wall or another perch just south of the nest. When incubation is in smooth operation Ave generally brings the kill to South Point, or their favourite look-out perch east of the eyrie, and entices Mavi off the nest with it. He may then relieve her for periods of about two hours, up to four hours on fine days, while Mavi does some hunting. He leaves the nest immediately she returns to it.

The incubation time is 44 days but the period is 47 days when two eggs are laid and the second-laid of these hatches.

CHAPTER SEVEN

Eaglet's birthday

As dawn broke cold and grey over the sea in the east, Mavi stirred in the darkness of the eyrie and turned towards Ave on his night perch opposite, momentarily lifting her oval blanket of brood feathers. Probably Ave could hear Eaglet's importunate cheeping inside his newly chipped and cracked egg-shell. Ave had expected these tidings that morning. Two days previously he had first heard Eaglet's sharp chirps and taps while relieving Mavi on the nest for his last turn of the season. He knew now that he must provide his yet unborn nestling's first meal of fresh, warm, tender dassie. One of a family of five little dassies, born a fortnight ago on the ridge below the eyrie and left undisturbed until now, would be suitable food for the chick's first meal and birthday feast, when the ceremony and ritual were over.

Ave launched himself into the dense, cold air and winged across to his watch post, trilling back over his shoulder to Eaglet and Mavi as he flew. He watched patiently for the sun to rise because its warm glow would, within two hours, lure the dassies from their den.

Far away down in the north-east, beyond the Tzitzikama, the sun sent light-laden pathfinders ahead to spread a rainbow cloak of strawberry and amber over the mist-covered shoulders of the dark mountains of dawn.

Ave kept his silent watch on the renascent day.

The sun climbed higher, behind the mountains. Above them the roan with strawberry flecks gave easy way to chestnut and chestnut to pure gold when the rising sun, eclipsed by Mount Formosa, spread half-orb and flooded the peak with diffused brilliance.

Back in their Eagle Hill Eyrie, Mavi played her matronly role. It was her duty to contain the nascent Eaglet in his chipped and split egg-shell until he was no longer wet around the ears. All round him in the oval shell was a fluid and even after he should succeed in sundering his tough skin embryo case and the shell his soft cream-coloured down would be soaked and matted and he would have to be kept warm for about seventy minutes before he was dry and allowed to see his brave new world and its sun, still hidden from him by his mother's warm belly and fully-feathered legs. She held him gently but firmly in his place there, in his shell with its irregularly chipped-out opening and fracture uppermost.

Eaglet battered more vigorously at his shell lining and wall with his beak and egg tooth. The fissure was allowed to open slowly, at Mavi's discretion as she released pressure from her legs on the shell's sides. When Eaglet finally broke it in two, she crushed one half with her foot and cradled Eaglet in the other. She would nurse him so until he was completely dry and warm and rested, and the sun would be shining on the nest. He was cradled at the precise moment of the crescent sun's appearance. Mavi had seen to it with experienced efficiency, as her movements showed.

In this manner was Eaglet born: not under a lucky star, as some men are, but at the auspicious moment of sunrise on a mid-July morning.

Ave had watched the natural ceremony in deep absorption and perhaps with feelings we men will never know. Now again on this festive day he turned to his duty.

His sharp eyes fixed on the clefted shelf, deep in cliff shadow, where the dassie family still awaited the sun's warming shafts. The sun rose higher and, when broad beams crossed their lintel, the whole family emerged behind their parents. Ave, rigid as a sphinx, watched them position themselves and, noting one move round the shelf out of view, marked it for his victim. Stealthily he moved back several paces on his rock, out of the dassies' vision. There were no warm currents rising yet and he had no need of them, as he had foreseen when planning this early morning swoop. He launched himself not towards the dassies but westward, towards the eyrie, against the gentle southwest breeze. Unseen by the dassies, he stooped swiftly down, wheeled sharply eastwards, swept round a buttress rock and swung up above the shelf with the victim in his talons.

On the ridge above he quickly decapitated and disembowelled it, and, clutching it in his right foot, circled up for straight flight to the nest, where he dropped it beside Mavi and left at once to hunt further afield. Mavi looked at it, approved, and drew it under her wing to keep the flesh warm, without disturbing Eaglet's dry-nursing.

Eaglet's birth ceremony had been no different from those of his ancestors, for Nature has always officiated. But this time a man with professed civilized proclivities had attended it.

The ceremony over, the man went away and returned with a young photographer who had learnt wisdom for he now knew that his best films and photographs of eagles were obtained when he respected their privacy and studied their ways and dislikes sympathetically. He was not interested in obtaining sensational pictures of angry, frustrated eagles whose fathers had been shot by trigger-happy boys with .22

rifles. Today he was to have his reward. He was privileged to put Eaglet's post-hatch and first meal on permanent record with cine and still cameras.

Three days earlier, on the morning of the 15th July at 11 a.m. we had filmed and photographed Mavi taking over incubation duties from Ave and seen that there was only one egg in the nest, which I recognised as the smaller, unmarked, second-laid egg of the original clutch of two. We had also photographed Ave later on the look-out perch east of the eyrie. Next morning I saw the egg again when Mavi stood up to stretch her legs and wings on the nest at 9.45 a.m. as soon as the nest was in shadow. On the morning of 17th July, Mavi and the exposed egg were again filmed on the nest. The pre-hatch had therefore also been recorded.

On Eaglet's birthday the camera was set up for filming at 9 a.m. Mavi was still dry-nursing Eaglet. Her sitting posture when nursing a chick is higher than the incubating close-cover position. At 9.29 a.m. she rose carefully and presented Eaglet to us beside his half-egg-shell, now of no more use. The proud mother, she looked for our admiration but all she got for her friendly thought was the cold, glassy stare of a purring cat from the cine camera. I opened the curtain wider for a moment to make amends. She allowed Eaglet to look around his home and enjoy the warm sun before it left the nest: then covered him again. He was not unlike a domestic fowl's yellow chick but his legs were big and feeble. We also saw the dassie with some red meat visible.

Eaglet's first feed

At 11.20 a.m. Mavi got up and, bending down, touched his beak with her own which she had previously rubbed on the meat. Instinctively he raised his big wobbly head on long neck and "chee-reep"-ed for food. Mavi placed the dassie conveniently and, with her foot on it, selected the best, firm red meat. She fed him pieces the size of minced meat and held them, one at a time, just above his head for him to stretch up and take from her beak. She went on for almost twenty minutes until his crop swelled to golf ball size on a body no bigger than a tennis ball. His crop can hold more than his belly can. For most of the meal he swallowed at the rate of one piece every three or four seconds. Whilst a downy chick he received two, and often three, such meals per day, all of fresh dassie. Mavi eats up all old remains. He was fed entirely by his mother until he was 42 days old. Only then did he really begin trying to tear meat from the carcass himself, whilst his mother held it down for him.

Nest cleaning

After Eaglet's first meal he was moved a short distance so that "crumbs" could be picked up and also the pieces of crushed egg-shell. The method of moving small nestlings, up to the age of about 40 days, is very gentle indeed. Mavi eases her three front toes (2nd to 4th) of both feet under the chick and shuffles slowly forward with the chick riding on them. To see a great, gaunt battle-axe of a mother eagle use her terrible, muscle-ringed and taloned toes, capable of vice-like grip, to move her tiny, helpless chick in this gentlest of ways makes one marvel.

Moving older nestling

In order to get Eaglet to move out of the way when he was older, on his 43rd day, Mavi placed her foot on his and increased pressure, making him draw back. She always ensures that his retreat will be in a safe direction towards the cliff and not the exposed side of the nest. As she places her foot on his she looks at him, in a meaningful way, and speaks. Her upper and lower mandibles open and close briskly in exhortation and her expression is severe and easily distinguished from expressions of encouragement to him, as when teaching to eat, for example. After the initial gentle pressure a firm tread and command sufficed. By then the toughening-up process in his education had begun.

Mother alone feeds and trains young nestling

Until Eaglet was 15 days old (1st August) all his nursing, care, protection, feeding and training was done by Mavi while Ave dropped food on the nest and usually left at once. On the 1st August Ave mounted guard on a rock above the nest whilst Mavi hunted. I first saw him on the nest when Eaglet was 25 days old (11th August). He always sleeps on his rock perch overlooking the nest at night. Mavi slept on the nest with Eaglet until ten days before he left the nest. She covered him under her breast at night, or on very cold days, until he was well-feathered.

Eaglet's first feathers

A Black Eaglet normally changes from the downy chick to the earliest fledgling stage in a week, starting from the 28th day of its life.

The following observation records tell the story of Eaglet's change from down to first feathers and, incidentally, raise other matters.

Sunday, 11th August (Eaglet 25 days old). A fine day after bad weather. I found Eaglet asleep on the nest at 2.00 p.m. with Ave also drowsing on the nest and the remains of a dassie beside him. They had all evidently had a good Sunday dinner, because their crops were very full, and I thought Mavi was probably getting some exercise. In fact she was hunting and brought back a big dassie which was on the nest early next morning. That afternoon the weather turned nasty again and heavy rain fell on the night of the 12–13th. She was laying in provisions. There had been heavy rain, and snow in the mountains, on the 8th–9th.

Eaglet is still a clean, downy chick.

Wednesday, 14th (Eaglet 28 days old). Eaglet's cream colour has become a little dirty in appearance but no feathers are visible through the down yet.

Thursday, 15th. This was a very disturbed day for the eagles and the chick was starved.

Saturday, 17th, afternoon. Eaglet was alone on the nest and cowered with head flat down, watching the hide when I opened the aperture. Fortunately Ave, who was overhead when I went to the hide, soon came to the nest, glanced up at the hide and quickly communicated re-assurance to Eaglet. The parents left at once to hunt. Eaglet has sooty patches showing through the down on wing shoulders and tips. I had never seen him cower, before this day, in fear of man, although he had learnt to crouch and cover meat with his body when gulls flew over the nest. On the 19th and 20th I watched Eaglet fed full meals by Mavi, each taking about half an hour and finishing at 12.07 p.m. and 11.35 a.m. respectively. Eaglet is still unable to feed himself but is being coaxed to make a start. He is through the first stage of plumage change, with narrow black fringes to wings and tail and patches appearing on mantle and wing shoulders. His chest is muddy coloured.

Wednesday, 21st August (Eaglet 35 days old). To hide 11.30 a.m. Strong, cold south-west wind. Eaglet sitting on half a big dassie carcass, hard up against the cliff wall out of wind. His primary and tail feathers have started growing out and feathers are just emerging through the down on mantle and scapulars. Chest is still muddy red. Mavi came to the nest at 11.33 a.m. and immediately began feeding Eaglet. During this meal she herself ate all the bones, including three leg bones complete, swallowing them whole. At the end of the feast when Eaglet's crop was very full and he had had more than enough, she gave him some small pieces of skin with fur to help digestion. She then turned the whole skin inside out and swallowed, taking two minutes to gulp it down and swinging it from side to side to stretch it as she forced it down into her crop, gasping with the effort. Not one hair of the dassie remained when the gorging was over by 12.07 p.m. Mavi then moved Eaglet into south-west corner of the nest, against the cliff, for protection from the wind, by pushing her front toes under him and shuffling forward with him. She then adjusted a stick to form a railing on the north-east side of Eaglet and stood close up to him shielding him with her breast from the very strong, cold south-west wind. I left them in their haven of rest when observation stopped at 12.15 p.m.

I estimated, purely on visual observation, that at the age of 15 days (1st August) Eaglet had grown to three times his size at birth and at the age of 35 days to at least seven times. My yardstick was a body that could be folded into a tennis ball with some liquid around it at birth.

CHAPTER EIGHT

Strangers at the nest

Unauthorized persons, usually fishermen, sometimes disturb the eagles when they have eggs or young in the nest, by walking over Eagle Hill, along the cliff and past the eyrie. On their approach the eagles ring above the eyrie and fly out to soar over them, returning to circle above the eyrie again. Mavi may then station herself on a high perch in view of the nest while Ave continues swooping in all directions.

Immediately the intruders are seen heading towards the eyrie, the eaglet on the nest is warned of danger. Its reaction is to lie flat on the nest, in the basin with its beak resting on the rim and facing outwards. It lies perfectly still, not blinking an eyelid, but perhaps a nictitating membrane occasionally. Its plumage, after the downy chick stage especially, blends with the background of sticks and faded leaves on the nest and it remains in this position until the parents indicate that the danger has passed.

The initial warning for downy chicks is a yelp but later on there appears to be no vocal communication. The young are made aware of danger, or of its passing, either by the parents' attitudes or by means of communication other than auditory, for the eaglet will sometimes relax tension when the parents are, apparently, out of its sight and hearing. Elephants and other animals give scent signals warning of danger. Black Eagles give the red, amber or green light by remote control. How they do so I am still investigating. Buffalo stamp their feet on the ground and Bushmen put their ears to it. Do Black Eagles use a wave length?

After downy nestlings have had danger-warning experience they will cower at once when I appear above the nesting cliff. Then one of the parents, usually Ave, will fly past leisurely above me and the nest, immediately dispelling all fear. Older nestlings soon learn to recognise men whom their parents regard as trustworthy, even though the men may change clothing and hats. I nearly always remember to greet them with open palm and arm stretched so that the palm will be seen.

Defence of home territory

When the Black Eagles are absent during midsummer one or two pairs of Jackal Buzzards (*Buteo rufofuscus*) invade their territory to hunt. Ave's first concern after his return early in February is to evict them. This he does by boisterous aerobatic display over his eyrie followed by sorties in their direction. The buzzards apparently concede his right to the area and leave without coming to blows. Protection of his food supplies is not Ave's only reason for driving them away. It is even more important that he should prevent them from nesting in the vicinity because Jackal Buzzards and Black Eagles are unable to share overlapping territories without fighting later, during the Jackal Buzzards' breeding season, when they are the aggressors and will attack the eagles whenever they pass their nests. For the same reason Ave has to prevent Cape Ravens from nesting on or near Eagle Hill. He ignores the Steppe Buzzards (*Buteo buteo vulpinus*) which enter his territory only as summer migrants from the Northern Hemisphere. African Marsh Harriers and Black-shouldered Kites, although resident, are tolerated. Peregrine and Lanner Falcons and immature Crowned Eagles, which have occasionally entered his territory for a day or so, are probably made aware that they are trespassing by Ave's display behaviour and leave without incident. Two Bateleurs have come singly to spy out the land and have not been challenged. They have kept away from the Black Eagles' nesting cliff, as have the Crowned Eagles which come to the three forested kloofs of the reserve. It is hoped that a pair of each will eventually come to settle in the reserve.

In July, 1965 a pair of Fish Eagles (*Haliaëtus vocifer*) appeared to be making an attempt to occupy the Black Eagles' alternative nest when Sixtyfive was a downy nestling in the main nest. It is just possible that they may have been the original owners of this nest or occupants during a Black Eagle interregnum. The nest was constructed on a bush growing on a ledge but the bush is now dead and the nest rests on the ledge. On the other hand, this south-facing, surf-lapped coastal cliff area would seem to be too remote from the lagoons and estuaries, some seven to ten miles away, which the Fish Eagles frequent for fishing.

When they came to the nesting cliff of Eagle Hill both Black Eagles engaged them in aerial combat. Mavi's tactics were to defend her nestling by holding the battle front between the Fish Eagles and the nests while Ave lashed at both opponents with savage pugnacity and eluded their co-ordinated attempts to close on him by wheeling or slipping under or over them in a swerving motion matched only by swifts and spinetails in the air or human skiing champions of the telemark. The scene of battle shifted eastward and over a high cliff top instead of the sea. Then Ave suddenly swooped down along the cliff edge and, taking full advantage of the rising currents there, shot up high over his opponents. Having gained this advantage he immediately stooped with terrific speed and raked one of the Fish Eagles across its back with his hind talons, plucking some small feathers. That was the end of the melee. The Fish Eagles turned inland, over a valley, and made off in the direction of Plettenberg Bay and their usual haunts whilst Ave returned at once to his eyrie with Mavi.

The Black Eagles had not provoked the fight but they had learned from experience that the best defence is offence. Ave had been merciful. He could easily have ripped into the Fish Eagle and wounded it mortally or decapitated it, as other eagles would have done with such an advantage: and the Fish Eagles knew it. Ave's intent was to prove his superiority. His constraint may have been inherent in his own good nature or it may have been instinctive. A quality of Black Eagle nobility is that they are not needlessly destructive of life.

Since the Battle of Eagle Hill in 1965 a Fish Eagle has come on reconnaissance in June or early July each year, flying high over Eagle Hill and passing rapidly eastwards, after giving a challenging cry but without waiting for it to be taken up by Ave who has merely watched from his favourite perch. A peace settlement between these two noble eagle families would bring victory for our nature conservation policy. I have seen Crowned and Fish Eagles sharing overlapping territories in the Albany District, and I still entertain hopes for Black and Fish Eagle compatibility here.

A pair of Rock Kestrels (*Falco tinnunculus rupicola*) have nested for the last two years (1967 and 1968) in a crevice on the west arm of the Black Eagles' eyrie cliff about seventy yards from the main nest and nearer to the sea, laying their eggs in September. The male Kestrel was a perfect nuisance to the eagles and especially to Ave, diving at his head many times in a day as he flew by. Later he and his mate all but caused Eaglet's death by drowning in the sea and eventually he paid for his aggression with his own life. Ave seemed quite unable to prevent the Kestrels from nesting so close for they were determined and quite undeterred by him. This indomitable little Kestrel, whose tribe I have known and admired for so long, was doing his duty defending his young family from possible danger in the only way he knew. I have had my cap snatched off my head by a Rock Kestrel when looking into a hole in a dead tree, where his brood was, and have had the same experience when climbing to a nest of the Red-breasted Sparrowhawk (*Accipiter rufiventris*).

After seeing the bold and persistent attacks made on Black Eagles by Kestrels, Peregrines and Jackal Buzzards when they are nesting nearby, I have often wondered to what extent, if any, the eagles are deterred from nesting at certain sites because of this nuisance; and the same applies to Cape Ravens. As the Black Eagles normally begin nesting earlier than their tormentors they are not troubled by them at a stage when they are most sensitive to interference of any kind.

The Black-backed Gulls (*Larus dominicanus*) sometimes mob the eagles at any season. Five or six gulls may be seen diving on Ave when he is perched on the cliff top. He watches but ignores them, although occasionally he has to duck his head to avoid a scratch or knock. I sometimes think the gulls are urging him to go hunting. At other times he seems to enjoy their frolics or perhaps he just feels that it is not worth wasting energy on chasing them away. But his mood changes shortly before Mavi is due to lay her eggs.

On one of these days he moved to a position above his nest and twelve paces away from my hide. The gulls followed and continued mobbing him much to his annoyance, as I could see. Then Ave coolly worked out his plan of attack. He feigned lassitude and the gulls grew bolder. He waited patiently until a gull came from behind and touched his head. In a flash he shot after it and it fled straight out to sea, keeping a level course so that Ave had to make wing with rapid beats, not having the advantage of altitude for a stoop. Ave was soon gaining on the gull and would have caught it, about three hundred yards out, had the gull not taken a perpendicular dive and plunged into the ocean. At that point Ave turned round and came straight back to his perch, looking very pleased with himself. If Black Eagles can chuckle, and recognise a human smile in return, it happened when he looked at me then, long and hard. His reasons for this show of force were many and good. The other gulls had vanished and kept away for many weeks whilst eggs and downy chicks were in need of protection. Ave had taught them a sharp lesson.

The mind of the Black Eagle

Black Eagle behaviour studies are fascinating but not easy to explain in ethological terms. We must avoid humanising eagles by implying that they have minds, thoughts, feelings or intelligence of human quality.

Their behaviour patterns include "inborn" reactions and also actions resulting from experience of life and environment, as do human ideas which are not innate as was once supposed.

We shall see that eagles are aware of human reactions, particularly when an act of theirs might bring trouble. This suggests a long natural association with mankind.

We shall also see that they appear to have feelings which they express in ways which we can recognise. But we can only speculate on how they feel and must not assume that they feel as we do.

When I record "intelligent" behaviour by my eagles I do not wish to imply that they use human intelligence. They do have what I call "awareness-wisdom" for coping with situations.

Free Black Eagles are quite undeniably capable of planning how to cope with a menace, to organise a hunt, sometimes using other birds to assist them, to forecast the weather and plan accordingly, and to deal with the problems of educating their young for life. Their young have to be taught such things as how to set about eating their food, how to use air currents and winds for flying and how to kill their prey. Intelligent application, with teacher-pupil understanding, is a *sine qua non* in the education of both man and other animals. Finally, Black Eagles are adapting their lives most intelligently in order to be able to coexist with man. Changes in the future are likely to be accelerated by human interference with their natural conditions. Those interested in animal psychology and bird behaviour cannot fail to find much of value in the study of the Black Eagle way of life today.

The Black Eagle fight for coexistence with man

Viewed from the land side, Eagle Hill rises up some five or six hundred feet from the valleys which hem round it. The hill is a sugar-loaf with a small, flat crown above a rim of rock. It is covered with macchia scrub, thinning on top. It stands in the south-west of the 400-acre nature reserve. All round the reserve, on the land side, are farms where stock and poultry range and tractors pull ploughs. Only half a mile to the north-west is a riding school and just beyond it the aerodrome. Twice daily Dakotas of the regular Cape Town to Port Elizabeth service come in to land and then take off, and quite often, when the wind so dictates, circle over the eyrie or just to lee of the cliffs. A busy main road runs parallel with the coastline, less than a mile north of Eagle Hill. Our Land Rover climbs up Eagle Hill itself, sometimes to a point only forty yards from the eyrie cliff, and the eagles take no more notice of it than they would of a Basuto pony.

The eagles accept this noisy, bustling, man-made world of today with equanimity, knowing they have got to live with it. At first they are very suspicious of further intrusions by strange things making strange noises. A week after Eaglet left his nest and was living on a terrace below, men in a ship began drilling for oil on the Continental Shelf, eventually striking gas forty miles out to sea, and helicopters, plying between the aerodrome and ship, came airscrewing monstrously and, in the view of the eagles, with menace and malice in its action of swinging outrageously low over their home cliff.

I saw Ave fly out to intercept the first one and, greatly fearing, rushed up to Eagle Hill to reassure them. There I found both the old birds sitting on their look-out post gravely watching the helicopter go out to sea and Eaglet warned and crouching below. I sat with studied unconcern on the cliff-rim directly above Eaglet and some thirty-five yards away from his parents. On its return trip, the helicopter again came directly towards us over the sea then began increasing altitude to pass over Eagle Hill. Ave watched the infernal spluttering aggressor restlessly and was of a mind to go after it again. He left his perch and circled round over Eaglet and Mavi but when the monster lifted and then whirled past us, he evidently felt that he had made his protest with good effect and dropped down beside Mavi. She had sat tight and tried to restrain him before he took off by speaking to him sternly. After the third day of the strange aircraft invasion the eagles were almost able to ignore them. Soon the helicopter service changed route to one passing east of Eagle Hill. The eagles must have seen the men in the cabin on the return trip, as I could. Then perhaps the strange thing passed muster for just another man-made contraption to be avoided.

Approaching eagles

It is not difficult to approach within a stone's throw of wild Black Eagles if one uses good methods. My most successful method of approaching eagles in the veld has been to note their position from a distance and then apply myself, with really serious concentration, to some other activitity such as flora field study, stopping to examine specimens; collecting firewood, if others do that in the area; admiring the scenery or casually netting butterflies. I stop frequently to look all about me and stare at the eagle no more and no less than I do at other objects. I do not walk directly towards it and I avoid sudden noises or movements, even if I have to allow a rare, fast butterfly to escape.

When near the eagle, I watch for any signs of restlessness and, if I feel that it will not allow me more latitude, I make the first retreat move by turning my back on it for a time and retracing a few steps. I might then try to get a little closer or wait for another day if the eagle is in the habit of using a regular perch. Eagles are creatures of habit and soon grow accustomed to men whom they feel they can trust and, believe me, they are excellent judges of human nature. If we wish to know them better we must first win their confidence. That will require patience for we must not make our intentions too obvious. In my shooting days I often saw more birds or game when I went out without the gun. The game seemed to have a way of knowing.

Frustrations and reactions at breeding time

Eaglet's birthday had been a very happy one: the best the eagle family had had for many a year; and men friends had shared it.

Two years previously (1966) their attempts to breed had

been utterly tragic. Despite all my efforts to prevent it, the eagles were badly disturbed at daybreak very shortly before Mavi was due to lay her first egg.

Ave on his night perch, immediately below the hide, yelped an urgent warning to Mavi and flew out to circle and watch from above. He had no need to. Mavi, less than sixty yards away from the hide and in the direct line of camera fire, could see and hear more. Terrified and angry as she was, she would not leave the nest, causing Ave more anxiety. She stayed and laid the egg in distress and panic at a time when, of all times, eagle mothers must have complete peace of mind. The disturbance continued. Then desperation drove her to atavistic cannibalism, an instinctive reaction almost as primitive as reproduction itself and aroused by fear and jealousy. She devoured her egg.

Possessiveness is deeply ingrained in Black Eagles: "What we have we hold and never waste." This must also have prompted her impulsive act, done with aquiline arrogance. If men were going to steal or destroy her egg, as she feared, she knew there was only one thing to do, there and then. What she did, not knowing it perhaps but instinctively, was to uncreate the egg in order to use its creative substance for re-creation through the natural process of metabolism in her body. She had not taken life, but she had destroyed its conception—only for a short time if all went well. By a provision of Nature her body would be able to produce another clutch of eggs after two or three weeks, provided the cause of her disturbed state was removed and her mind was at ease. That was not destined to happen in this case. Meanwhile the second egg due to be laid three days later was not laid, but assimilated by her body.

The reason for the eagles not laying again that year is not proven but I believe that it was suspicious fear and perhaps loathing of Hide I after their terrible experience. Before the tragedy, they had accepted the hide, and the fact that men made and sometimes used it, and had carried on normally with their nest repairs and courtship. Because they knew me, they always received me well and without alarm. It was strangers they feared, particularly if unaccompanied by me.

They began to repair their alternate nest, which is not overlooked by Hide I, nor seen from it, although situated almost directly under it and seventy feet below in the middle of the eyrie crescent. Mating began again but with less courtship play and continued over a period of nearly three months, even without preliminaries later. They were left undisturbed but eggs were never laid, not in the nest at any rate. Their state was one of fruitless frustration. The next year the eggs were laid very late and in the alternative nest.

While no one would wish to contest the rights of accredited wildlife photographers, whose art has been of inestimable value in wildlife protection and in many other fields, we must ensure that we do not interfere with the undeniable rights of wild creatures, legally protected by man, to be spared avoidable disturbance at breeding time, and especially just before and after eggs are laid or young are born.

It is grievous to see photographs of eaglets cowering on the nest. It means that they fear man because they have been warned against him by their parents. It means that the photographer is not recording the family's natural behaviour as he should be doing. It is a sad reflection on attitudes to wildlife when such photographs are published. Even worse are pictures of newly-hatched chicks, still wet after their emergence from the egg because their mothers have been frightened off the nest when dry-nursing them. Eaglets which have had fear-of-man experience will always regard man as an enemy, unless their outlook has subsequently been changed by men with better understanding of them. Yet it is an ill wind that blows nobody any good for, paradoxically, the well-disposed but thoughtless persons responsible have at least warned the young eaglets that men are dangerous animals. It is up to man to demonstrate his friendly intentions by an understanding approach. But we must first ensure that all men will be friendly, lest we deceive the eagles by betraying them into the hands of the treacherous. The case is overwhelming for educating people to the eagles' need for our friendship; to the fact that they are well-deserving of it; and in methods of winning their confidence.

My eagles are always aware when anyone is in a hide, no matter how quiet and well-behaved the visitor is or whether or not his entry was observed. They are more wary of some strangers than others, as may be clearly seen from their scrutiny. Sharp bangs or rattling doors and loud talking annoy them and sudden movements or lighting cigarettes in front of the observation window distract them. Camera lenses should not protrude. Attention to such details will help to ensure that the eagles are filmed carrying on their home life peacefully.

The rhythm of life

"There is a tide in the affairs of men which taken at the flood leads on to fortune." There is a rhythm in the lives of Black Eagles which left undisturbed leads on to well-being. That there is this rhythm I am now quite certain after fifty years of study but my understanding of it is still incomplete. The principle may well be established in future that when Nature is allowed to take her course untrammelled and untampered with by man the animal lives involved will be enabled to develop to genuine fullness.

My observations have established beyond all doubt that young Black Eagles, Secretary Birds and Wattled Cranes (*Bugeranus carunculatus*) bred unnaturally late in the season received inadequate education and training for efficient living when they must become independent of their parents. They are not born fools but victims of upset rhythm. Similar problems arise when a fledgling is unnested before it is due to leave its nest. The rhythm persists in periods as short as days and hours. There are clear indications, for example, that daybreak to sunrise is the rhythmic hour for Black Eagles' (and others') eggs to be laid and chicks to be hatched. This cannot be proved or disproved by those who cause the mother eagle responsible for these events anxiety resulting in rhythmic disturbance and estrangement.

No true natural historian (or any other historian for that matter) interferes with the course of events or the way of life he is recording. If the eagles are disturbed by an observer's interference his results are bogus. He is behaving like a successful Canute and he is no naturalist.

The freedom of a naturalist is not licence to behave wilfully. It is freedom to do what he ought to want to do, having due regard for the welfare of his subject. There is enough vandalism to legislate against without having to deal with disturbance.

My eagles' rhythm of life during Eaglet's year at home swung naturally most of the time, more so than during the previous year. Some of the observations recorded in this book were subsequently checked by two eminent naturalists who observed aspects of these eagles' lives in my company.

CHAPTER NINE

The eagles at home

It is sometimes stated that the remains of dassie or other kills are found in Black Eagles' nests or below them. My eagles leave no remains whatsoever for the simple reason that they consume the complete kill brought to the nest.

All bones, including the skull and backbone which are neatly dissected, the skin and fur as well as the flesh are swallowed. Sometimes, but not always, the entrails of a dassie are removed and left for the gulls. The carcass is generally brought to the nest without the head which is eaten at the kill or on a rock perch on the eyrie cliff. I have searched the surroundings of these perches, regularly for nearly four years now, and never found any skulls, bones or skins. A guinea-fowl brought to the nest was plucked on the nest and all the feathers thrown to the wind, in swift sweeps of beak and neck during the plucking operation. But two large primary feathers, together with the breast bone, were carefully kept as educational toys. Those are the only items of any kill which have ever remained on the nest for long when Eaglet was a nestling. Even the smallest chips are eaten up during the nest cleaning after a meal.

My Black Eagles have a traditional and systematic way of opening up a dassie carcass and removing the flesh, sinews and bones, starting from an opening at the neck and peeling the skin back from the body, and from each leg in turn, so that when they have finished the skin is intact.

Each of the three eaglets bred here have been carefully taught by Mavi to skin and eat a dassie in precisely the same way. They watch the method carefully as they are fed pieces of meat which get larger as they grow older. Later on they try themselves and are encouraged and helped by Mavi. She eats some herself and when a bone has been separated and cleaned she swallows it or sets it aside for the older nestlings to swallow at the end of the meal. The long leg bones can be observed moving down into the crop until they pack away, goodness knows where and how. Finally the skin is turned inside out and rolled up and stretched between feet and beak. One end of the roll is swallowed and after each gulp the eagle's head is swung vigorously from side to side to prevent a traffic jam. She gasps for breath, her mouth and throat crammed full. Eventually, after between two and fourteen minutes on different occasions, the whole skin disappears into the crop and, after a minute or two, she has recovered from the discomfort. Her crop is now full to capacity. It bulges perceptibly and her breast feathers stick out because her crop's skin is so stretched.

At various times during the last three years I have observed Black Eagles swallowing complete dassies after dismembering them in the manner described.

And now I have a witness in the person of one of our most distinguished naturalists. On the 15th October, 1968 Dr. Cecily Niven spent the morning in Hide I with me because Eaglet was due to fly from the nest any day and I wanted her to observe the role of the parents.

Unfortunately the morning turned wet and cold and the parents made no effort to encourage Eaglet to fly on that day. Instead Mavi shared a big dassie with him on the nest, the last food he would have before he flew two days later. Between them they demolished the dassie, including the bones, and Eaglet got a little skin but Mavi the lion's share. On this occasion she took fourteen minutes to get it down, her slowest and best on record. She never lets me down.

My record reads:

"15.10.68. Eaglet 90 days old. In Hide I with Cecily Niven, 8.40 a.m. to 1.15 p.m. Wet day with cold west wind. Mavi on nest watching Eaglet eating dassie and later had some herself. They ate most of the dassie except saddle and back legs and skin. Each swallowed a front leg bone whole. Then Mavi left but came back again after about fifteen minutes and the feasting was resumed. Mavi then swallowed the skin taking 14 minutes. Both their crops were very full indeed by midday. Cecily clearly hears Eaglet's cheep-cheeping or chirrup-chirruping. She counted eleven continuous notes on one occasion and sixteen on another. She also saw Mavi speaking to Eaglet but could not hear her. Ave came over once; then sat on East Perch. I told Cecily about the guinea-fowl and the hare."

Earlier I had pointed out to her the two guinea-fowl primary feathers still on the nest. Afterwards Mrs. Niven commented, "I shall never forget the experience of watching that mother bird 'talking' to her child and swallowing the dassie skin."

Our observation on the Black Eagle's eating habits seems to have escaped many modern observers but probably not South Africa's early naturalists who named the Black Eagle *Aquila vulturina*, as recorded in "Birds of South Africa" by Stark and Sclater, Vol. III, page 292.

Home life peaceful and happy

Black Eagle home life is peaceful and affectionate. The parents show much forbearance when their young are fractious and patience when training them. They are never gruff or grumpy and they never scold or quarrel about food. When a kill is brought to the nest, older nestlings are allowed to take possession of it and mantle over it. Long before the parent arrives with food, and is still out of sight, the eaglet jumps up expectantly and ready to grab it. If the eaglet has difficulty in skinning and eating, the mother approaches him slowly and indicates that she will feed him by touching his beak with hers. He then gives over the food to her and while she feeds him she teaches him how to set about skinning and dividing the carcass. This is always methodically done.

After feeding, the mother usually preens the nestling until it is old enough to preen itself everywhere.

Body hygiene

Black Eagles do not bathe themselves in water, although they get wet in rain and snow. Nor do they sand-bath. Their method of cleaning themselves externally is to preen their feathers and clean their skins with their beaks and sometimes they give themselves a good shake after preening. Feathers are drawn through the two mandibles and old ones, or down in the case of chicks, are removed. This is particularly necessary when downy chicks grow into fledglings and when moulting takes place. As we shall note, chicks are taught by their mothers to preen themselves and take to the operation like ducks to water. The head and throat are the only parts they are unable to reach with their beaks. These are cleaned by scratching or combing with the toes. Beaks are also cleaned by rubbing on sticks or rocks or with the toes.

Internal cleanliness is of great importance because the intake of food, including fur or fine feathers and bone, into crop and stomach is high and digestion quick. The waste material must therefore be expelled frequently. Before Eaglet had his second feed on the afternoon of his birthday, Mavi moved him on her toes to the edge of the nest and, using her beak, massaged his abdomen, with the desired result—a strong clean discharge. It is remarkable how soon very young chicks learn to go to the rim of the nest, lean forward and do this themselves, nearly always down wind if possible. Food is often withheld until they have done so. The rate of defecation varies with the rate of food intake as a general rule.

While we are on this subject it is worth noting that both male and female invariably expel faeces before the sexual act. Black Eagles always defecate before they fly off on any activity after resting. This is to be expected with their high intake of food and rate of metabolism. It is a useful sign that a bird is about to fly for photographers and stalkers.

Weather-forecasting

My Black Eagles are most economical and efficient housekeepers because they are remarkably sagacious weather-prophets. They appear, in practice, to follow three principles: first and foremost the food must be fresh and preferably young dassie is fed to young nestlings; food is never wasted because the parents eat up all remains; and supplies are laid in before bad weather makes hunting difficult or impossible.

When the weather is settled they kill food as needed but when bad weather is expected they begin to make provision four or three days ahead. For the first three or two days their intake of food is increased and their crops are kept filled to capacity. On the last day before the stormy period, spare food is stored up on the nest and sometimes also in a niche about ten feet from it.

During the bad weather they shelter from the storm, Mavi and Eaglet on the nest, usually in the south-west corner under the overhang, and Ave in the niche round the corner. They lie-up and rest, if necessary for a day or two without eating. Then the parents eat sparingly but feed their eaglet well.

Many times have I noted and tested their ability to forecast weather accurately several days ahead and to plan accordingly for housekeeping. They are never wrong although often two or three days ahead of the Weather Bureau forecasts I receive on my radio at home in the evening. Plettenberg Bay weather is not easy to forecast because it lies between two weather cycles—Sir Lowry Pass to Plettenberg Bay and Plettenberg Bay to Port Alfred—but their forecasts are always reliable.

Just as remarkable is the fact that when the young bird is induced to make its first flight from the nest the wind is from the south-west. Any other wind would be unsafe because of the dangers of drowning in the sea or striking cliff faces and damaging feathers.

Full crops before bad weather

22.8.68. Eaglet 36 days old. Brown feathers showing through down on breast and more sooty patches on upper body; otherwise plumage as on 21.8.68. 11.00 a.m. To Hide. Cold south-west wind and cloud since noon yesterday. Ave overhead and then to East Perch as I pass by. 11.15 a.m. Mavi alights on nest with fresh young dassie. Ave had flown above and behind her guarding the rear as she carried it. Her crop is full and the head and front portion of the dassie are missing. She speaks to Eaglet, standing on the dassie. She waits for Eaglet to defecate then starts feeding him. He squats on his tarsi. She faces him with both feet on the dassie carcass and hind and inside (1st and 2nd) talons gripping it. She tears meat off the carcass by an upward pull with her beak. The white V on her shoulders is visible but her wings cover the white patch on her back and the primaries cross over her tail and are as long as the tail when she is erect. She feeds pieces about every three seconds, pausing at the top of her tug for Eaglet to take them in his beak. After 30 minutes his crop is full and he lies down. He is still unable to feed himself but had tried to tear some pieces off with Mavi's help and encouragement. Feeding in the sun but by noon both were in the shade. 12.10 p.m. Eaglet restless. Mavi moves up and he settles down to sleep under her breast. 12.20 p.m. Mavi left the nest and circled overhead with Ave. 12.25 p.m. She returned with a bar stick and placed it north of the nest basin. She then deepened the basin which is still kept well-lined with fresh rooikrans and pine twigs. She placed the small remains of the dassie in it and moved Eaglet on her feet and settled him on top of the dassie skin. This is a precaution: the kestrels started building their nest in the crevice beyond the eagles' nest today. Gulls pass all the time. 12.40 p.m. Eaglet asleep in shade in basin. Parents have gone hunting again. Observation stops.

23.8.68. Heavy rain and wind from south-west.

Eaglet mantling and preening

29.8.68. Eaglet 43 days old. Main plumage change is more dark brown feathers with lighter fringes on his wing shoulders. 11.40 a.m. to hide. Warm day. No wind. Sea very calm. Eaglet on nest in full sunshine, hot and panting. Crop empty. Ave on East Perch. 11.50 a.m. Shadow begins to reach nest. Eaglet still in sun panting. 11.55 a.m. Nest half in shadow. Eaglet happier and looking about.

12.00 noon. Nest in shadow. Eaglet lively and expectant.

12.05 p.m. Mavi comes to nest with most of fairly young dassie except head and front legs but all the skin hanging loose. As she reaches landing stage at south-east of nest, Eaglet scrambles forward and mantles over dassie, his wings spread out like a fan, tips on floor of nest and stubby tail also; head down at Mavi's feet and his own feet on red exposed meat. Mavi stands on carcass and lets him try to tear strips off it. He succeeds in tearing away three loose pieces (perhaps previously prepared by Mavi).

12.10–12.55 p.m. = $\frac{3}{4}$ hour. Mavi fed Eaglet who sat back on his tarsi as usual. Eaglet defecated twice, half way through meal and at end of it. On each occasion he stood up on his toes on rim of nest, tail pointing north, bent forward and discharged cleanly and far. At 12.55 his crop was very full and Mavi stopped feeding him but he wanted more. She stood on dassie remains—saddle, back legs and all skin—and let him have a go at it. He soon got tired and gave up. He pushed his head and back under her breast and rubbed them against her feathers. She preened his head and back only today. Previously she had preened everywhere since feathers had started growing.

1.00 p.m. Mavi flew off and joined Ave on East Perch. Eaglet settled down next to dassie remains facing south-west with his left wing spread a little, showing the dark markings, and the other tightly closed and resting against the dassie skin but covering all exposed meat. This wing and the dassie skin together now looked exactly like his left wing and twigs on nest made a good camouflage. Gulls are flying over all the time. The kestrels are flying to and from their nest and mated on their perch opposite it. A big dassie came out and sunned itself on the cliff top immediately above the eagles' nest and seven feet back from the cliff rim. Fifteen Red-winged Starlings fed round the nest and my hide. Rock pigeon come and go and sit about.

1.10 p.m. Mavi returned to the nest with a pine twig and put it on south rim of nest. Eaglet stood up. She placed both her feet on top of his and put some weight on them. Eaglet pulled back under overhang. Mavi excavated a hole in the basin floor and ate some old bone chips she found there. Then she left and joined Ave and soon went off hunting with him. Eaglet pulled the dassie remains into the hole with his beak. He stood on his toes and walked round the basin unsteadily, sometimes falling on his tarsi but up again, until he reached the new pine twig. This gave him a firm level stand. He bent down to reach the dassie meat with his beak but fell forward, his legs too weak. After some pulling at the dassie skin whilst squatting on his tarsi he gave up and lay down beside it. Two ravens came circling round over the nest with their claws down. They finally decided that discretion was the better part of valour and went away grudgingly. They could probably see Ave coming but I could not. Eaglet then lay down on top of the dassie in the basin completely hiding it. He began preening himself all over and gave himself the best cleaning I had seen so far. Much fluffy down drifted away on the breeze. He soon got sleepy and was asleep when observation stopped at 2.00 p.m. Mantling, preening himself fully, except head and nape, and protecting food from gulls, ravens and kestrels: he was growing up.

Eaglet tests his wings

5.9.68. Eaglet 50 days old. He has grown much since 29.8.68. When he stretched his right wing full length it was about three times the width of his shoulders. A black band runs along the ends of the primaries and secondaries. Wing coverts are mottled. Wing shoulders and mantle dark brown. There is still some creamy down on his back but it is mostly dark or tawny. His toes have changed from skin colour to yellow and talons are black. Beak and ceres yellow. Front centre of upper mandible has a black shield. Black spots on ear-coverts. Crown and forehead lightly streaked with tawny. Nape still downy. Eyes very black and set in dark sockets. Middle of chest heavily streaked with black feathers. When wings are raised the back is white and there is a whitish V across shoulders. White trousers cover leg (thigh or tibia) and foot (tarsus) down to toes. Belly white. Tail very short and slightly mottled and barred.

11.10 a.m. To hide. Warm, quiet day with light north-east wind. Sea calm.

Three fairly fresh pine sprigs on nest and many faded pine and rooikrans twigs well trampled down on outer rim as well as centre which is now a saucer where basin was before. Nest and Eaglet half in shade. A gull worrying Eaglet by flying past two feet from nest. He jumps up each time it passes. He is quite active on his toes now and moves about easily. He lies down half in sun and yawns. He is feeling the heat.

11.42. He moves into shade. His crop is half full. Ave on East Perch and Mavi on another perch east of nest. Both can see nest but had ignored the gull. They have been there since 11.00 a.m. Eaglet watches starlings and kestrels flying about with intense interest. The kestrels mate again.

12 noon. Both parents have disappeared. Nest in full shade. Eaglet stretches right wing and leg backwards full stretch together standing on left foot. 12.07 p.m. he faces cliff standing in middle of nest and flaps his wings four times. They are heavy because the quills are very full of blood and will be almost until the wings are fully developed. The effect is to lift him slightly backwards and he ends up on the exposed rim of the nest. He walks back to rock face and turns round facing east.

12.15–2.03 p.m. He moved about on nest preening himself most of the time. He can now reach every part of his body except his head and nape. He pays special attention to neck and chest. He watches all passing birds including a mollymauk out to sea. He is now very alert and quite active. He exercises his wings again gently. This is the first day I saw him test his wings. He defecated cleanly twice during the three hours of observation, which ended at 2.07 p.m. Observation was resumed from 5.10 to 5.30 p.m. Eaglet walking about on nest and preening and looking round from rim of nest. Parents not back from hunting.

CHAPTER TEN

Black Friday

The 6th of September was a day of fasting and danger for Eaglet in unsettled weather and a hard one for his parents. It began with a hot north-east berg wind and muggy sea air and ended in a black sou'wester which brought heavy rain that night and next day.

9.30 a.m. Eaglet's crop looks empty and there is no food on the nest. Now he is feeling the heat. Ave is on guard duty from East Perch and something is worrying him for he keeps patrolling the cliffs on each side of the eyrie. Eaglet watches him and frequently calls to him complainingly between stretching and preening and watching the other birds. Both Red-winged and European starlings have begun making their nests between the big sticks near the base of the eagles' nest. A cormorant is sitting on eggs below and others are building nests. The male kestrel comes to and fro feeding his mate on their nest. Several Rock Pigeons sit about near Eaglet. Black and Alpine swifts sweep past to their nesting crevices and wheel round screaming excitedly.

10.25 a.m. Eaglet stands in the middle of the nest and begins flapping his wings, slowly rotating round the nest, his toes sometimes lifting off it. Then he stretches, first right wing and leg, then left wing and leg full stretch backwards and sideways. 10.47 a.m. He gets tired and sleepy and lies down, obviously suffering from the heat. Three Rock Pigeons settle on a sloping rock just above him and he watches them. 10.55 a.m. A Peregrine Falcon sweeps round South Point. The pigeons scatter but the falcon comes on and circles round Eaglet, three feet above him. It begins a second and lower circle then suddenly dashes off as Ave sweeps past Eaglet and chases the falcon round South Point at incredible speed. The male kestrel dives over South Point following them. A minute later Ave comes back over the cliff, circles round Eaglet, who speaks to him, and perches on the end of South Point. Eaglet calls him and he answers but keeps looking west along the cliffs. The male kestrel has followed him back and mobs him a little. He ignores it. Ave must have anticipated the peregrine and flown a high pitch somewhere behind my hide and above the south slope of Eagle Hill. He could never have swooped on the peregrine so soon and fast from East Perch or any other stationary position.

11.40. Eaglet now in shade and much more lively. He defecates and then stands on rim watching cormorants and a spur-winged goose which comes round South Point and flies east.

12.00 noon. Nest all in shade. Ave goes eastwards and Eaglet watches him, crying; then remains sitting on his tarsi on rim of nest looking north-east. 12.20. Observation stops. 5.00 p.m. To hide. Very overcast with strong south-west wind. Both parents hunting east of eyrie. Eaglet sheltering under overhang and crying. Crop still empty. He watches two starlings two feet away on dwarf wall. It is getting dark.

5.25. He stands up facing east and crying and after five minutes lies down still crying. Then gets up and watches gulls and pigeons.

5.35. Female kestrel perches on South Point; then male comes and they mate.

5.36. Eaglet suddenly lies down flat on the nest facing east and silent.

5.40. A Cape Raven swoops in rapidly from the east, making straight for Eaglet. It passes the nest, wheels round five feet above it with claws down, and, with a raucous "kraaak", begins to drop. Suddenly it swings away and, losing height rapidly, disappears round the base of South Point, as Ave, followed by Mavi, reach the point and speed round it at a higher level. The eagles soon return over the point. Eaglet jumps up calling and watches them keenly as they swoop along the cliffs eastward again.

5.47. Eaglet backs to rim of nest and defecates to south-east. Then returns under overhang.

5.48. Both parents now hunting, sweep along the cliff, Ave below rim and Mavi just above, and pass on westwards to Still Bay. It is dark and foggy. A pigeon almost settles on Eaglet and both get a big surprise. The swifts are silent in their crevices below my hide, clinging to the rough rock vertically as they roost.

6.01 p.m. Mavi alights on the nest rim with Eaglet waiting to meet her on landing stage and getting in her way in his eagerness. He mantles at her feet. Nothing! He is complaining now, then fretful, then sulky. Mavi gently bundles him into the south-west corner with her breast. Then she jumps on the dwarf wall and looks out anxiously eastwards where Ave is now hunting. Eaglet faces her whimpering. She takes no notice. Eaglet turns his face to the cliff and sulks.

6.10. Mavi jumps down and moves up to Eaglet sheltering him and speaking to him. He is soothed and soon begins snuggling under her. Her back faces my hide and I catch a glimpse of Eaglet's head between her legs when she bends forward. It is dark when I leave the hide at 6.15 p.m. How did Eaglet get the long distance warning of the raven's attack? Did the raven's mate get away with their kill out east while the raven attacked Eaglet to draw the eagles away?

What we have we hold

8.9.68. Eaglet 53 days old. Observation from 4.40 to 6.22 p.m. High tide, rough sea and haze. Mavi on East Perch. Eaglet glanced up when I opened the hide window; then took no more notice of me. He is lying down and on the south rim of the nest is a portion of dassie. His crop is very full. There are more black feathers on his ear-coverts and cheeks. 5.04 p.m. A gull flies over, spots dassie remains and calls other gulls.

5.05 p.m. Mavi lands on nest as Eaglet rushes forward and mantles over the remains, snarling at her. This goes on for a minute. Mavi's crown feathers rise as she puts her left foot firmly on dassie and stands over Eaglet. She gently pushes him back with her shoulder. He tries to recover it, pushing his head between her legs. She preens his back and gives him a tweak. He draws back sharply. After a few minutes he walks round behind Mavi and defecates; then tries to recover the dassie again. She pushes him back with her chest. She eats some "crumbs" picked up where Eaglet had been lying. A European Starling perches three feet away from her head. Black saw-wing swallows have joined the swifts wheeling round the cliffs. They will nest in our sandbanks later. Other birds living with the eagles carry on as usual.

5.30 p.m. Mavi scratches her beak with a toe casually. Sun setting behind cloud. The gulls fly over nest and investigate. Mavi moves partly off dassie so that they can see it and she ignores them as they mill round.

5.40 p.m. Eaglet suddenly springs forward and mantles over dassie under Mavi. She resists a little; then gives way. Eaglet takes possession very pleased with himself and sits over dassie. Mavi also very pleased. They sit on peacefully for a time.

5.55 p.m. and 6.10 p.m. Twice more Mavi takes possession of dassie remains and makes Eaglet battle to recover them.

6.22 p.m. Mother and nestling with dassie under him snuggle together in usual sleeping position. It is dark now. This was an extremely well-taught lesson. Black Eagle families never fight amongst themselves for food.

A change of diet.

9.9.68. Eaglet 54 days old. Warm, still day. To hide 11.00 a.m. Eaglet mostly in shade under overhang and lying with wings parted and back exposed. He preens most of the time because all the fluffy down must give way to feathers and they must be given clean air to be healthy. He watches gulls and other birds passing, pausing in his preening. The dassie remains he held over last night have disappeared. Had them for breakfast?

11.25. Eaglet springs forward and calls "chink-chink", two metallic notes. Mavi approaches gliding low along cliff face and swings up to the landing stage. Eaglet mantles and grabs a lizard and swallows it whole. Almost immediately Mavi looks round and takes off flying round South Point and on to Still Bay. Eaglet watches her, smacking his mandibles. He looks more like a vulture than an eaglet, with his short rounded tail, white at tip and ringed with black and buff bands; and mottled brown wings with black end bands. But he has long white trousers and yellow toes. His head and mantle are tawny.

There is nothing more of special interest to record for this day. Man has, I believe, never before recorded the fact that Black Eagles feed lizards to their nestlings. They have probably done so from times long before man existed. I have seen four lizards and one frog fed to eaglets here when they were between 39 and 67 days old: also other small creatures swallowed whole so quickly that it was impossible to identify them.

The eagles all had full crops by evening and very full ones by 11.00 a.m. on the 11th September. By midday a very strong, cold sou'wester was blowing and by 1.30 p.m. the skies were black and rain threatening.

More plumage changes

13.9.68. Eaglet 58 days old. Fine, cool day with west breeze. I spent the hour before noon in the hide in order to check Eaglet's plumage changes. When I started walking up Eagle Hill Ave left East Perch and came circling over me and then the valley, gaining altitude rapidly as he ringed. A raven joined him and started dipping and touching him from above, but he did not seem to mind. He occasionally flapped two or three strokes on sharp turns. When really high he set his wings back and soared against the breeze at high speed over Eagle Hill and Still Bay and on to the sea cliffs beyond. The raven was left standing.

Eaglet was lying mainly in shade under the overhang where many sticks had been removed to give more room. The nest itself is flatter and wider and well matted with faded twigs. Eaglet spends much of his time spreading his wings and legs out on these as he lies about. His crop is very full indeed and he has a well defined line of black feathers down the middle of chest and breast and some black markings on the outside of his white thighs. His wings are now fully feathered and black but the feathers on his wing shoulders are fringed with tawny borders. His mantle and head are now completely covered with russet feathers, slightly darker on the mantle and contrasting with dark cheeks. He exercised his long wings full-stretch and dancing on tip-toes first 5 flaps, then 9, then 6.

16.9.68. Head and mantle are a richer russet now. Chest and breast white with black streaks like an adult Martial Eagle but the centre line is darker with black arrows. No down in front now.

A herbal relief for engorgement

16.9.68. Eaglet 61 days old. Calm day and sea with light variable winds early and strong, cold south-west wind after 11.00 a.m.

On this day Eaglet was grossly over-fed; then given a cure which he is unlikely ever to forget; then the feasting continued and the cure was again administered. The feast began at 8.40 a.m. and finished at 4.00 p.m. when Eaglet, assisted by Mavi, who ate mostly bones and skin, had consumed a very big dassie. The dassie was killed early in Rondebosch Rivier Kloof and carried to the nest whole by Mavi, guarded by Ave.

Soon after 10.00 a.m. Eaglet's crop was uncomfortably full and he lay down gasping and panting. His head and shoulders were shaded by the overhang and Mavi used her wing to shade his body but I could see his back heaving. He was too full and too hot.

At 10.40 a.m. he got up and defecated. He had last done so at 9.50 a.m. Now he was ready for the second course and Mavi fed him red meat and some entrails. At 11.05 Eaglet stood up because he could not swallow any more sitting on his tarsi. Now Mavi fed him some pieces of skin and fur to help digestion and five minutes later she was plying him with meat again. Now in the cool cliff shadow the valiant trencherman did his best to get outside the meat which Mavi kept on proffering. Feeding stopped at 11.20 a.m. when Eaglet had to defecate. He had been grossly over-fed again and he lay down in anguish.

A minute later Mavi dropped off the nest and was joined at once by Ave from East Perch. Together they flew the Multi-V way in the now fairly strong and cold sou'wester to Still Bay. Meanwhile Eaglet, his crop bulging like a balloon, lay gasping on the half finished dassie. Mavi had eaten the big bones but he had been given the smaller ones. At 11.37

a.m. he stood up and defecated once more; then lay down again sorely distressed.

Mavi returned to the landing stage at 11.46 a.m. and spoke to Eaglet. She preened his head and neck and made him move by treading on his foot. She excavated a fairly deep hole in the floor of the nest near the centre with her beak. Then she flew off and picked a fresh spray from a bush growing on the cliff rim north-east of my hide and brought it back to the nest. She ate some of the green leaves and, holding the spray in her beak, induced Eaglet to eat some. Then she placed the sprig in the hole and got Eaglet to lie over it. His wings and legs rested on its edges and his crop was suspended over the void. Eaglet had defecated again at 11.58 a.m. and the leafy twig had been brought exactly at noon. I noted carefully where she picked it.

While Eaglet lay gasping and panting Mavi watched him carefully. At 12.47 p.m. he was a little better and began playing with the twig, using his beak, and ate some more leaves. Mavi registered approval. Three minutes later he got up and sat on the rim of the nest. His panting spasms were fewer and he was clearly recovering from his excessive feeding. Heat could be discounted as a contributory factor from 11.00 onwards because the wind had turned cold. At 1.00 p.m. Eaglet moved back and again lay over the hole with his weight off his crop. He looked more comfortable now. Mavi sat quietly watching him, like a wise old owl with her eyes half closed, and began crooning to him. By 1.10 p.m. they were both relaxed in siesta.

At 1.27 p.m. Eaglet got up and defecated. Three minutes later he was walking round flapping his wings until he put a foot in the hole and fell heavily. He picked himself up, still very clumsy on his feet, and facing the cliff, began flapping his wings again. His left wing struck the overhang and he was knocked backwards, narrowly escaping a fall over the north rim of the nest. Mavi, almost in panic, as her eyes showed, made a forward movement then stopped, allowing Eaglet to learn his own lesson from his own frightening indiscretion. He scrambled back on all fours and lay over the hole again. He was much better now, thanks to the herbal remedy and the exercise. But he seemed to be a little intoxicated.

At 2.00 p.m. Mavi started eating again. Eaglet jumped up, put his feet on the dassie and started tearing some for himself. Mavi fed him too. He is now well and active as he sits on his tarsi with talons on the dassie. Mavi gives him some skin and fur.

At 2.15 p.m. Ave was back on East Perch with another dassie, as Mavi and I noticed. She went on feeding Eaglet from the dassie on the nest and ate some herself. Eaglet is now tucking into red meat with a breakfast appetite. By 2.40 p.m. he is literally fed right up to the gullet. His neck sticks out in front like a pipe above a crop swollen out like a balloon. A minute later he walks round and defecates northwards. Mavi interrupts her skinning of the dassie to watch him and approve. He returns and lies down over the hole, his feet on each side of it and facing east.

2.44 p.m. Mavi leaves the nest, circles round, joins Ave who greets her, touching her beak, and they sit together on East Perch. 3.14 p.m. Mavi returns to the nest and sits next to Eaglet. He is still lying over the hole with wings half spread and very distressed. She consoles him, talking quietly, and touches his beak; then pulls out the leafy twig and holds it in front of him. He pecks at the leaves and perhaps swallows some. (I cannot be sure.) His face is grim and he stares with glazy eyes. Every now and then he hiccups and shakes his head violently. His left wing covers the dassie hind leg and skin which remain. She lifts it, pulls away the dassie leg, tears off and eats some skin; then gives him some small pieces too.

3.40 p.m. The wind has dropped but the sea is boisterous and sky cloudy. Ave is eating his dassie. Eaglet gets up and defecates. Then he moves round the nest and puts a foot on the dassie leg. Mavi tears the skin off and Eaglet sits down and pecks at the flesh. Mavi gives him some small pieces and eats some herself. I can see the white bone now.

4.01 p.m. A gull dives over three or four times, a foot over their heads. Eaglet stands straight up almost leaning backwards to balance his crop. He pulls the dassie remains into the hole and lies over it.

4.06 p.m. Mavi hops up on the dwarf wall south of the nest and settles down to rest. They have both had all the food they can take but Eaglet is guarding his breakfast. Ave is still busy eating when observation stops at 4.30 p.m.

Eaglet was deliberately and repeatedly crammed with food to crop capacity all day long in order that he could be taught this herbal relief for engorgement. He will never forget its appearance, taste, smell and curative value or fail to associate it with relief from acute discomfort. The lesson was extremely well taught as any educationist would agree. It is interesting to note that no green twigs of any kind had been brought to the nest for some days previous to this. I had wondered why. My hide notes for this day cover seven pages of foolscap, closely written because so many interesting points arose.

Black Eagles are not gluttons. They make the best use of food when it is plentiful and they know the meaning of starvation.

A day of rest

17.9.68. Observation from 11.00 a.m. to 3.00 p.m.

All very quiet after yesterday's feasting. Ave on guard and sitting mainly on East Perch all day but patrolling at hourly intervals when the male kestrel usually mobbed him. He sometimes ducked his head or did a side skid when the kestrel swooped on him but took it all in good part. Mavi away all day. Taking her turn at hunting? No dassie remains on nest so Eaglet had his breakfast. The nest rim had been built up six inches high on the north side where Eaglet nearly fell over yesterday. Mavi had not forgotten.

CHAPTER ELEVEN

The training of an eagle

On the 18th September Eaglet was 63 days old. His tail was now 3–4 inches long, black with light brown fringe and upper tail-coverts. The central patch on breast and cheek and ear-covert patches were darker and bigger.

Observation from 2.30 to 3.30 p.m. Sun appearing through cloud after a dull morning with strong sou'wester. White horses gone and sea calm. No sign of parents. At noon Ave had soared fast and high over Eagle Hill going from east to west. The barrier sticks are still on the north rim of the nest but a big, heavy stick wedged firmly into the south rim near the dwarf rock wall has been exposed. There are two fresh rooikrans twigs in the middle of the nest and Eaglet is lying on them.

He notices at once when I open the shutter and calls; then takes no more notice of me unless I put my hand out. He is keenly interested in all the birds around him and follows their movements with his eyes. An Orange-breasted Sunbird, feeding on heath flowers on the cliff top in a shaft of sunlight, absorbed his attention for four minutes.

2.40 p.m. He yawns and, lying almost on his back, stretches his legs in turn; then stands up and stretches left wing and leg together full stretch rearwards. He defecates south-east. Now he preens his rather empty crop. He almost reaches his throat, his flexible neck doubled up. Now he faces the cliff and exercises his wings, first two half beats, then seven fullwinged but they fade at the end of the stroke.

2.50 to 3.00 p.m. He begins tugging the big exposed stick and scrapes the bark off it with his beak. Now he lifts his right foot, grabs it and pulls hard. This goes on for some minutes. After a time he tires and looks at the other birds again.

3.15 p.m. He tugs at the stick again with his right foot. Then climbs upon it and looks over the dwarf wall for the first time, snaking his neck about as he takes in new scenery. He watches a pair of rock pigeon making love on a ledge beyond. In a self-taught lesson he has learnt to perch on a single stick. Which parent arranged this new educational toy for him to find? He jumps down and hurts his toes on some sharp up-ended sticks below the perch. He squats on his tarsi and feels his toes with his beak. He flexes his toes. He gets up, walks round on smooth sticks and climbs upon his perch stick again. He has learnt to perch all by himself; but there was a highly skilled teacher in the background.

The snake stick

19.9.68. Eaglet 64 days old. Strong west wind and heavy cloud after rain earlier. Huge rollers. Observation 10.30 a.m. to 5.55 p.m.

Mavi lost no time after Eaglet's bright showing yesterday. She brought another stick, shaped like a puff-adder ready to strike, dropped it on the nest and left at once, shortly before 11.00 a.m. Before this Eaglet had been standing on the stick perch watching the two Red-winged starlings which had a nest in the base of the eagles' nest and were now sitting on the dwarf wall two feet away. When Mavi came he noticed that she carried a stick and did not mantle. Now he lies down preening beside the stick.

11.02 a.m. He gets up and starts playing with the stick. It curls round him when he grips its head. It does all sorts of things to him in retaliation for his attacks with beak or claws. He is fascinated by it and goes on playing with it for nearly two hours giving all his muscles exercise and learning the uses and strength of his beak and talons.

1.01 p.m. Both parents swoop past, round South Point and fly the Multi-V way against the strong west wind to Rondebosch Rivier Kloof. There two ravens appear and they each chase one, disappearing in the shadows in steep dives. They have gone to hunt. Meanwhile Eaglet has some good wing flaps, facing the cliff; plays with his snake stick again and rests, always preening.

4.00 p.m. Sun out now but nest in shade.

4.14 p.m. Mavi lands with a headless dassie, Ave guarding her from five screaming gulls. Eaglet rushes to meet her, mantles, grabs it and pulls it onto the nest. Mavi sits on the perch watching and speaking to him. He stands on the 4 lb. carcass, tears off meat and eats. Now he removes skin and eats fast. He is really coping very well indeed by himself today. This seems to confirm that Mavi herself fed him for much of the time three days ago in order to over-gorge him. Ave flies overhead, having got rid of the gulls, and sits on East Perch. The two on the nest looked up and watched him passing.

4.45 p.m. Eaglet is having some difficulty skinning. Mavi encourages him. He succeeds and goes on eating meat. Five minutes later he is in difficulty again and looks at Mavi. She comes forward. He crouches over carcass. She moves slowly round and he gives over. She begins feeding him and he watches how she manages the difficult joints.

5.00 p.m. Sun shining in west and wind moderating.

5.14 p.m. Ave comes over nest high, drops his feet in signal, and goes hunting at Still Bay. Mavi continues to show

Eaglet how to divide a leg bone, and so on, and he watches her methods ever so carefully.

5.30 p.m. Now she has extracted intestine. She shows him how to expel the contents by pulling a length through her beak and shaking it. She eats some; then gives him a cleaned piece. Now she gives him a piece which has not been cleaned. He shakes his head and disgorges it. She cleans it and gives it back to him. He has got the idea now and cleans some for himself. All the leg bones have been shared between them and swallowed whole, with little difficulty by Eaglet after watching Mavi dispose of one or two.

5.40 p.m. Only skin left now. It is whole, excluding head and tail and a few small strips Eaglet has had. Mavi turns the skin inside out and stretches it, then starts swallowing it whole. She gulps and gulps and swings it from side to side so that it does not bunch in her throat. Her eyes are closed with the effort. Eventually she gets it all down. It has taken her four minutes and another to recover. Eaglet has watched her with intense interest, imitating her actions in sympathy like a man watching a high jumper clear the bar. I am sure Mavi spun it out for him. Now he picks up some scraps from the nest floor. Mavi jumps up on the dwarf wall and Eaglet stands in the middle of the nest looking east. Both their crops are full. It is amazing how much they can put away, bones, skin and all.

5.55 p.m. Observation stops. Ave not back but probably having supper at Still Bay. He will bring some for Eaglet's breakfast.

Ave on nest duty

Although observation was fully maintained from 20th to 25th September when the eagles' activities at the nest and their hunting methods, using ravens as their stooges, were recorded, space will not allow more than brief reference here.

Eaglet's first flight from nest (*Photo: W. T. Miller*)

Eaglet's plumage became darker generally and his tail and wings grew faster and received plenty of exercise and preening. He and Mavi were photographed resting between 4.00 and 5.00 p.m. in the afternoon of the 25th September, together with the male kestrel flying near the hide.

One important point must be mentioned. At 3.15 p.m. on the 23rd September, Ave brought a 2–3 lb. dassie to the nest and remained on the nest until Eaglet had finished it. Hitherto Mavi had always carried the kill to the nest and stayed to assist Eaglet with his feeding. On this day she had come to the nest at 12.30 p.m. empty clawed. Eaglet, who had been crying with hunger, had jumped up long before I was aware of her approach from the west, where Eaglet could not have seen her, and was bitterly disappointed to find that she had no food. They had both bent down, their heads close together, and spoken to each other, beaks opening and closing once per second. This conversation went on for seven minutes. At 12.37 p.m. Mavi had flown off and Eaglet then looked happy. It had been a dull cloudy day with a strong sou'wester but the sun had come out just before Mavi arrived and the wind soon began to drop. Then turned into a south-easter. Now Eaglet stood under the overhang relaxed until shortly before Ave arrived from the east with the dassie, when he came forward ready to mantle and pull it under the overhang. Had Mavi told him to be patient and had all this been arranged? I felt certain it had.

Ave made no attempt to assist Eaglet with his feeding and did not watch him as closely as Mavi does, though he did speak to him encouragingly.

Eaglet managed well by himself and ate the whole dassie, except for the head which Ave had eaten before bringing it, as I could see by the blood on his beak and face. This was the first time I had seen Eaglet consume a whole dassie, including all the bones and skin, by himself. And Ave had let him sort out all the difficulties of dissection on his own. When Eaglet had finished at 3.40 p.m. he came forward and stood proudly in front of Ave to show him. Ave looked pleased; then left and Eaglet stood watching him hunt along the cliffs eastwards for three minutes until he had to turn round and defecate north-east. He had to again at 4.20 p.m. At 4.05 p.m. Mavi, who was hunting with Ave, dropped down, like a falling leaf against the now strong south-east wind, into the nest and stayed only one minute waiting for Ave to manoeuvre an attack; then supported him by showing herself to the dassies on Still Bay east cliff. They killed a dassie at 4.45 p.m. as the gulls told me. As I walked down Eagle Hill track two coveys of Cape Francolin and a Grysbuck were there feeding quietly. The Eagles often circle over them but never disturb them. They could easily have killed one that afternoon and on many other days.

The eagles bring home a guinea-fowl and a hare

26.9.68. Sunny day with sou'wester.

On this day both a guinea-fowl (*Numida meleagris coronata*) and a Cape Hare (*Lepus capensis*) were brought to the nest; the first I had seen in over three years of observation. Hitherto nothing but dassies, a few lizards, a frog and some unidentified small creatures had been fed to eaglets here. There was more than one reason for the change of fare on this day, as I was to discover. I shall give only the salient points from seven foolscap pages of recording during over six hours of observation.

As I walked up Eagle Hill shortly after 10 a.m. I saw that Ave was not on East Perch where he usually sits until the air warms at this hour to create currents, when he often comes overhead to greet me before going hunting. It was my eldest son's birthday and my thoughts were with him and his family and about his important scientific research work. Bill

Miller and I had noticed that the eagles had quite full crops yesterday afternoon so that early hunting seemed unnecessary.

On opening the hide shutter I was surprised to find both Ave and Mavi at the nest staring at me with anxious concern. They usually glance up and take no more notice of me and this was the first time I had seen them at the nest together since Eaglet was born. The reason quickly became apparent. Eaglet was standing on a huge guinea-fowl, eating it. Having recovered from my surprises and realised why they were watching me so closely, I bent forward so that they could see my face easily, greeted them and turned to watch some giant petrels out at sea. When I next looked the eagles were at ease.

Although we had failed to establish guinea-fowl in the nature reserve, so near the sea, there were plenty of wild ones inland and the eagles had every right to them. Moreover, there would be times in Eaglet's life, on a journey for example, when dassies were not available and he must be taught now—he was 71 days old—to recognise them as food and how to eat them. This was clearly in the minds of the parents because, after the meal, two of the old bird's primary feathers and its breastbone, which had given Eaglet trouble, were kept on the nest as toys for him. Dassies he knew well for he could watch the families that lives on the slopes below his home.

At 10.25 a.m. Mavi spoke to Ave whereupon he shook himself and flew off eastwards, spiralled, returned over the nest and made for Rondebosch Rivier Kloof in the northwest. He was going to hunt again as I could see from his purposeful action.

Now Eaglet has trouble with the primary feathers and looks at Mavi who encourages him and he goes on extracting them. An hour later when he had got to removing the breastbone Mavi watched him closely and then persuaded him to let her feed him while he watched the method of dealing with bird bones. But she also had another reason. She began picking out some small, dark, round objects, holding them in the tip of her beak for Eaglet to see and then flicking them far away with a sharp swing of the head. What could they be? Guinea-fowl do not have round faeces.

By 11.50 a.m. Eaglet's crop was so full that he had to stand up straight and twist his neck about in an effort to contain the contents. He sat on his tarsi and watched Mavi pluck some small feathers, in rapid sweeps of her beak, and cast them to the winds. She tore off a piece of skin with small feathers and set it aside. Eaglet had eaten the wing bones and many smaller ones. Now she cleaned both drumsticks and swallowed them herself with considerable effort; then the thigh bones.

At 12.11 p.m. they both stopped, looked at each other and then up. A minute later Ave appeared, held wind over the nest, spoke to Mavi and returned north-westwards. Feeding went on and Eaglet was again fed to the point of acute distress. Mavi then went off to the spot where she had previously obtained the herbal remedy for over-feeding but, instead of bringing a leafy twig as Eaglet was obviously hoping, she came back to the nest and fed him the piece of guinea-fowl skin with fine feathers which she had set aside earlier. This was the right remedy for guinea-fowl-itis and, like dock leaf for nettle, honey for bee stings, it was handy. Mavi taught him this and her cure was effective. By 1.15 p.m. Eaglet was a new eagle, exercising his wings and preening.

Meanwhile Mavi had gone off again and was sitting on East Perch. Forty minutes later she returned and, between them, they devoured the remains of the guinea-fowl and cleared most of the feathers away except the two primaries which Mavi tucked into the sticks in the south-west corner of the nest and the breastbone which she hid under the north-east rim. Then they rested.

At 3.30 p.m. Mavi flew eastwards and five minutes later she was on her way back carrying something big in her left foot with Ave as rearguard. They both land on the nest and sit looking at me hard again. They have brought the backbone, without head or ribs, and the hindquarters and legs of a big Cape Hare. The tail is missing but I can identify it from the white fur inside the thighs because the carcass is on its back. The legs stick out stiffly over the nest rim. Rigor mortis has set in and Mavi sits between the open legs staring at me. Ave on the dwarf wall is also studying me. Eaglet is under the overhang scared of the skeleton. I show them my joy. Now I know what those round balls were that Mavi picked out of the guinea-fowl. They were AAA shot. There has been a shoot. The eagles have found the dead and wounded and lost game. They are bringing windfalls home. But they are afraid of the consequences because these wild creatures were shot by a man on his farm. That evening I check with my farmer friends and the Police. Yes, there was a shoot yesterday in the upper Piesang Valley about two miles north-west of the eyrie. Man did the killing. The eagles may have heard the shooting and watched. They fetched the guinea-fowl last evening or early today. Later Ave had found the hare, come and told Mavi, gone back and eaten all he could while the others ate the guinea-fowl, for there was blood on his beak; brought the remains to a shelf near East Perch, then fetched Mavi.

And it was necessary to lay in provisions. After one day there was heavy rain, soon to be followed by more. They were keeping it for a rainy day. The hare was not eaten that night: they all had very full crops.

We are told that eagles and animals cannot think: that they are unable to put two and two together like we can. I wonder? We have all watched our dog's behaviour when he picks up a dead thing that stinks to us but is perfume for him. He brings it to the doorstep and watches our reactions, begging us to let him enjoy it, just for this once, but knowing very well that he will be chastised. That is how my eagles regarded me, a man, on this day. They knew the risk of bringing home the guinea-fowl and the hare: the danger of reprisals.

I have been an educationalist all my life. I am now learning how my eagles think and discovering new and sometimes better applications of our educational theory from their practice of it.

Eaglet's last 21 *days as a nestling*

The eagles were photographed at the nest on the 27th and 29th September, the 28th being a rainy day. On the 3rd and 17th October, Eaglet's last day on the nest, Bill Miller photographed them with me, but unfortunately had to leave at 12.15 p.m. on the last day, three hours before Eaglet flew. On the 15th October Dr. Cecily Niven spent the wet morning in the hide with me, as already recorded. I kept the eagles under close observation for the whole period beginning 30th September and saw Eaglet leave the nest. It will be necessary now to select only major points of interest arising during this period.

Most noticeable were the daily changes of plumage as the eaglet became father to the eagle. At the end of the process he was a Black Eagle with black feathers and white feathers where they should be but in many places these feathers were fringed with various shades of rufous and browns producing a tawny or mottled result. His cheeks and chest were jet black, crown and mantle rich rufous but mantle darker than crown, back and legs whitish and mottled and other parts tawny. He had broad black bands on wings and tail and the famous Black Eagle whitish windows at the base of his primaries. His wing shoulders and fringes and tail fringe were buffy.

Most noticeable also at this stage was the acceleration of tail and wing growth at relatively varying paces:

30.9.68. "His wings grow longer daily but his once stubby tail is also growing very fast now. When the wings are folded over the tail, their extremities are two inches shorter than the tail fringe."

11.10.68. "Tail very long now—about four inches longer than wing tips."

16.10.68. "Tips of folded wings still two inches short of tail fringe." When the fledgling becomes a fully grown juvenile or immature young eagle its wing and tail tips lie level, as in adults when sitting in normal posture. Eaglet attained this wing development on the 18th December, 1968 when he was five months (154 days) old and two months after he left the nest. It is extremely important to remember this when we observe the young bird learning to fly. Illustrations showing adult Black Eagles sitting erect with wing tips shorter than tail are incorrect for live specimens seen in the field.

From the 30th September to the 15th October inclusive large fresh dassies were brought to the nest daily, except on the 3rd (unless after 5 p.m.), after a night of heavy rain, and the 10th, a day and night of rain with gale force winds. On each of these occasions the eagles had correctly forecast bad weather and filled their crops in good time. On the 16th and 17th Eaglet was starved. Early next day he was fed where he had landed on the terrace below the nest. Mavi continued to spend between two and six hours a day on the nest, sometimes shading Eaglet from the sun and giving him final lessons in the art of dismembering dassie carcasses. She let him do more and more of the work. Late on the afternoon of the 11th, after the weather had improved, he was given his first dassie whole, complete with head, and left to get on with it. Mavi had already shared another with Ave and looked tired as she watched Eaglet with eyes half closed. They had killed the dassie about 5 p.m. and she had to carry it in strong wind. When Eaglet had had his fill she always ate up the remains, usually one or two big leg bones and most of the skin, which she always swallowed whole.

After October the 7th Mavi went back to her night perch next to Ave's, leaving Eaglet to sleep alone on the nest so that he could get used to it because he would soon have to spend several nights alone on a sentinel rock at the base of the cliff until he could fly well enough to use his own night perch near his parents. Meanwhile Mavi continued bringing fresh green twigs to the nest.

Eaglet became more and more active on the nest, exercising his claws and beak frequently by using his collection of toy sticks and the guinea-fowl breastbone and primaries. When one of these feathers blew over the cliff in a southeaster on the 11th, Mavi brought it back and tucked it so deeply into the sticks beyond Eaglet's big stick perch that Eaglet had to tug away several sticks to recover it. In so doing he exposed a small rock perch below the dwarf wall south of the nest and about eighteen inches from it. That became his favourite perch for his last six days on the nest and was to be the last he ever used on it.

His wing exercises became frequent, strenuous and positively dangerous. His wings stretched out about a foot beyond the north and south rims of the nest, his wing-span being greater than the nest diameter, and he was unable to use his wings when facing north or south because of obstruction from the cliff and overhang. His wing beats carried him a foot or more above the nest as he bounced and danced upon it. He was only waiting for Nature's permission to begin really flying. The rich blood that had filled his wing and tail quills and nourished them during their rapid growth had made them heavy and rather soft. Now it was steadily receding and would continue to do so until the wings and tail gained the right proportion for early flight, namely tail to be about two inches longer than folded wings. Then the blood would withdraw from the hollow quill centres altogether, except for their pinion bases, and the wings and tail would become dry and hard and strong and very light with their hollow quills. That happened to Eaglet on the 17th October when he was 92 days old. On that day his wings were "hard summed" and he was ready to fly: but he was still only equipped for early, restricted flight.

CHAPTER TWELVE

Winning his wings

On the 16th October Eaglet was alone all day on the nest and unfed, and yet he was not lonely and not hungry. His crop had been filled to capacity at noon on the previous day and was still quite full late on this day.

He never cried as he does when hungry but frequently answered and called to Mavi who was sitting on the sloping ledge at the base of the nest about seven feet lower and three feet more in a northerly direction. Ave, who was sitting on East Perch at 10 a.m. had disappeared. I could not see or hear Mavi from the hide. Indeed, I have never heard her or Ave when they were obviously speaking whilst I hear Eaglet distinctly. That experience was shared by Dr. Cecily Niven and Bill Miller whose hearing is better than mine. This was a very good day for listening. The sea was quiet and the day fine with light variable winds tending to south and becoming south-east after 4 p.m.

In order to be able to see Mavi, Eaglet pulled away sticks from the north-east rim of the nest and disposed of them under the overhang. He did this several times during the day, speaking to Mavi all the time. When the nest was in full shade soon after noon, he scrambled far down the now sloping north-east edge and raised his wings, obviously considering flying but stopped and watched the swifts, starlings and many white butterflies on eastern migration flying past, following each one until it disappeared from his view. When a young gull flew by with a strip of white cloth about a foot long trailing from its foot he twisted his neck round almost full circle as he watched it takings the bends along the cliffs.

He exercised his wings vigorously very many times this day and sat on his stick and rock perches as well as the sloping eastern nest edge. He preened himself thoroughly and often and watched the downy fluff drift away.

It was clear that he wanted to fly now and was being encouraged in the idea by Mavi because a spell of very bad weather was on the way and she knew from experience and her rhythm controlled instinct that he was ready. But neither she nor Nature intended that he should fly before his allotted 92 days as a nestling. This was the last day of preparation. Eaglet was being conditioned. When I left the hide at 4.40 p.m. I knew for certain that he would fly next day and that evening telephoned Bill Miller, urging him to come with his camera.

When the camera was set up at 9.30 a.m. on the 17th October the west wind was rising and became steadily stronger and a little more southerly until nearly 4 p.m. An hour later it had moderated and by 6 p.m. it stopped. The day was cloudy with some drizzle at midday but at 3 o'clock the wind was perfect for a safe landing away from the sea.

Eaglet was resting earlier and his parents were not visible from the hide. Later Bill obtained excellent photographs of him exercising his long and now light wings and sitting on his south-end rock perch. After Bill left at 12.15 p.m. Eaglet flapped his wings more strongly than ever, always facing the cliff and resting on his rock perch between hard spells of exercise.

At about 2.30 p.m. I became aware that the parents were

nearby and that Eaglet was answering them but I could neither see nor hear them. By flying low along the cliff face they are able to reach many perches below and round the rock projection beyond the nest unseen from Hide I.

At 3.15 p.m. Eaglet tried to jump with open wings from his rock perch to the dwarf wall, eighteen inches away to the south and about a foot higher. His right wing struck the overhang and the wind, from which he had been sheltered by the dwarf wall, lifted him up and over the south-east edge of the nest, at the same time turning him right about.

He screamed and in a flash Mavi was under him and Ave in front as he recovered balance and flopped down towards the terrace at the base of the cliff in the northern arm of the eyrie crescent. I dashed out of the hide and flung myself down on the cliff rim behind it to see Eaglet safe on the succulent carpet far below and both parents next to him. Immediately the parents flew up to the nest together, inspected it thoroughly as though they could not believe what had happened: then flew round South Point and back over Eaglet to settle on East Perch. At 3.30 p.m. I moved back to the cliff rim below Hide II where I could watch Eaglet and ten minutes later the parents, satisfied that he was safe, went hunting.

Before Eaglet's fall I believe Mavi was sitting in the niche just round the corner from the dwarf wall so that she could immediately swoop under him from the sea side and that she was calling to him to come to the dwarf wall at the time. Ave must have been somewhere ready to guide Eaglet down to the soft terraced slope at the cliff base. That the event had been expected, if not planned by the parents, I have no doubt because Eaglet was starved for two days to make him light and restless.

They had played a more active and direct part in making Sixtyseven leave the alternative nest. She had been born much later than Eaglet, and was literally pushed out of the nest on the 20th November in a favourable wind. During the previous week Ave had himself pulled sticks away from the rim of the nest leaving a stick perch exposed to which he had edged the eaglet and induced it to use the perch. Then the parents had demonstrated take-offs and glides from a ledge nearby. After more demonstrations on the last day Ave got behind and back to back with Sixtyseven and pushed her over with his tail. She was urged, encouraged and accompanied by her parents on her early flapping flights during the next week and by the 1st December was flying well alone and took off without persuasion or fear. In her case the parents were obviously anxious to force the pace of her training because she was born so late in the season and, being a female, was stronger and developed more rapidly than the male eaglets. I have reason to believe that her nestling period was 93 days and possibly 94 and it was obvious that she did not want to leave the nest and had to be induced to do so. Once launched she did well. The drop from the alternative nest is much shorter and less dangerous than from the main nest, provided the wind is right.

Reports of young birds of prey leaving their nests independently of their parents, or in the absence of uninterested parents, always surprise me. Accidents can, of course, happen and the observer may sometimes cause them.

Eaglet's new home

Eaglet's safe landing when he left the nest had been on the sward near the one big isolated rock at the base of the middle section of the eyrie cliff. It stands four feet from the cliff wall and is about ten feet long and eight feet wide and high. It is sheltered from the west winds and behind it is protection from the east winds. Men and baboons never enter this secluded area. This became Eaglet's new home for a time as his parents intended, but not for the first night.

After the parents went hunting on the previous evening I watched Eaglet scramble up the sloping sward, then up a ridge east of the sentinal rock to settle on a ledge where ridge joins cliff. There he spent the night. The parents had not returned when I left at 6 o'clock.

Next morning a fairly strong sou'wester was blowing when I found Eaglet in almost the same position as the previous night. At 9.35 a.m. the parents came from the west with a very big dassie, flew high over Eaglet with it, then turned and came back low along the cliffs and, passing close to Eaglet, landed on the sward below the sentinel rock. Mavi remained there with it and Ave on a cliff perch above her and near me, calling Eaglet. He soon took off, against the south-west wind gaining lift, and got in 60 yards of flapping flight. Having landed clumsily on the sward beyond Mavi, he scrambled back and, mantling, began tearing into the dassie. After a time, when he grew tired, Mavi helped him and fed him very much. By 11.20 a.m. he had had his fill, including some skin and bones, and Mavi finished the remains. Ave had stayed on guard all the time, keeping a watch all round but unconcerned about me on the cliff rim above him.

Eaglet now began sleeping on the sentinel rock and using it as shelter in bad weather. The 19th and 20th October were wet days with a strong sou'wester and next day there was a gale force south-easter. After the 28th October Eaglet slept on his own night perch with his parents but they continued to bring him dassies to the sward below the sentinel rock until the 20th November, 1968 when he was 126 days old. He began hunting with his parents on the 21st November and after the 3rd December sometimes on his own and sometimes with them.

Dangerous days

These early days of freedom from the restrictions of the nest were dangerous ones for Eaglet. Adverse winds could cause his drowning in the sea as could gulls and kestrels by mobbing him and baboons would destroy him if they could. I did my best to ensure that he was not harmed by them or men inside the wildlife sanctuary created for their safety. But I saw him narrowly escape death on two occasions.

On the 21st October, four days after Eaglet had left the nest, he was lifted over the cliff rim in the strong south-easter and dropped down to shelter on the lee side of a thicket of stunted rooikrans bushes nine yards behind and north-west of Hide I. He allowed me to pass and enter the hide and as he seemed safe I left him there at 11 o'clock. When I returned at lunch time I saw immediately that something was wrong. Ave was stooping and wheeling over the thicket and, upon seeing me, came swiftly to meet me just overhead and back over the thicket. As I ran to the spot four baboons broke from under the thicket and fled with Ave diving on them in the open scrub. He then returned and soon he and Eaglet dropped over the cliff, leaning sideways against the wind, and Eaglet was soon safely in the shelter of the sentinel rock below. I found baboon spoor where Eaglet had been and one of his secondary flight feathers. This suggests that he had had a very narrow escape from the baboons. The loss of the feather was serious for a young eagle in his early flying days; yet I was unable to detect a gap in his remiges.

Late in the afternoon of the 24th October, a week after Eaglet left the nest, I watched him flying about the eyrie cliff and noted that he was coping quite well with the sou' wester and the fact that when settled his colouring blended well with the rocks: also that his wings were now scarcely an inch shorter than his tail when sitting.

Then he flew up and settled on the cliff near the kestrels'

nest. Both kestrels immediately attacked him, tearing at his head and face with their claws, and, when he dropped off the perch, pressed home their attack. Eaglet flapped down the cliff face and into the sea. A wave washed him under the arch in the cliff foot and he disappeared from view.

Ave, who had been watching, flew down to the sea cave mouth and then back to me. I knew that Eaglet must be in grave danger but there was nothing I could do. Even with two men and a rope it would take half an hour to reach the spot and by then Eaglet would be drowned. The only gesture I could make was to move further along the cliff rim and show my concern. When a huge wave broke and flooded the cave I thought that it had done for Eaglet.

Miraculously he came out in the backwash and first a rock pigeon and then Ave went to a sloping shelf east of the cave. There Eaglet saw them and flopped ashore on the next wave. Ave walked quickly up the slope and Eaglet managed to scramble after him before another wave reached him. Ave kept going, using his wings only to jump over obstacles and so led Eaglet to the sentinel rock. He had narrowly escaped drowning and for the next four days kept to the sward round the sentinel rock. There he scrambled about and played with sticks which he pounced upon, dragged with one foot and savaged with his beak. And he slept on the sentinel rock at night.

Ave never forgave the kestrels. At noon on the 11th December he came from the east and dropped low over me when I was standing beside our lower dam. In his claws was a male Rock Kestrel, as I could see from the victim's slaty grey head. He went on to the eyrie cliff with it. By this time the kestrel family had dispersed.

Return to the nest

After these terrifying experiences Eaglet was disinclined

to leave the safety of the lower eyrie crescent and his parents did not force him to fly although they continually demonstrated flying to him. By bringing his food to the sentinel rock they made him fly to it from any point where he happened to be at the time. This gave him some idea of wind cheating and air control.

In the afternoon of the 29th October the parents took measures to restore his confidence and this is how they did it.

Mavi spoke to him on the sward, where he had fed four hours earlier, and then flew up to the main nest and sat on the dwarf wall beside it calling him to her. After much chirriping and argument he mounted the sentinel rock and eventually flew up the wing-beating-way to South Point. There Ave came to him at once and, after circling round, glided past the nest and on eastwards along the cliff rim, greeting me with his feet down as he passed. Eaglet watched him while Mavi kept calling.

Then Eaglet dropped down smoothly on steady, levelled wings into the nest. Very pleased with himself now, he had a good look all round the nest inspecting every corner and was very excited at being back after his absence of twelve days. Meanwhile Mavi pretended to take no notice of him but she was watching him carefully, as I saw. Then he began to pull sticks away from the east side of the nest, just as he had done the day before he left it, and he stood there watching the swifts fly by. There were no white butterflies on eastern migration today but the visions he had had of flying things must have been recalled by the familiar scene and his urge to fly was born anew.

He turned round and ran across the nest to his rock perch. Now Mavi spoke again, pressing herself against the cliff to leave room for Eaglet on the east end of the dwarf wall. This time he jumped with wings closed and made the wall easily.

Mavi spoke again, dropped off the wall and sailed round

the crescent and along the rim of the north cliff to swing up and settle on East Perch. Eaglet had shadowed her all the way, sailing past me just as gracefully. He had learnt to ride the cliff-top air currents and they were wonderful for it was a warm, sunny day with just enough east breeze to buoy him.

One need not be a child psychologist or educationalist to realise that these eagle parents could not have found or planned a better way of restoring Eaglet's confidence. Their technique was perfect and they selected an ideal day for putting it into practice. What is still more remarkable is that their wisdom conquered prejudice. Once a young Black Eagle leaves his nest he never returns to it; nor do the parents until the next breeding season. On this day the parents knew they must return. Eaglet had fallen from the nest and on evil days. Now they were making amends in order to give him a fair start in a life which made heavy demands on flying ability.

New horizons

From East Perch Eaglet's horizons were spread wide open. For the first time in his life he saw green hills and mountain ranges and what young Black Eagle can fail to answer the call of the mountains? What stronger inducement to use his magnificent but as yet latent powers of mobility? That evening he surveyed his new world until the sun set behind the deep ashy blue Outeniqua Mountains. There were many silvery linings in the clear skies round the peaks and piebald clouds rode above on wings of flamingo pink. Small wonder that from this day onwards East Perch on the cliff-top became Eaglet's, as well as his parents', most favoured. After the sun had set at the end of a wonderful day Eaglet's parents took him to his night perch beside their own.

Learning to fly

Next afternoon at 3 o'clock Eaglet flew with both parents

and was given a really good lesson in cliff-top flying, returning twice to East Perch for rests. From this day he was given daily lessons, sometimes in quite rough weather, and made steady progress. Later on training in flying and hunting were combined.

When Eaglet left the nest he was unable to fly strongly for physical reasons as well as lack of experience. We noted that his wings were not fully grown, being about two inches shorter than his tail when folded over his back in a sitting posture. Not until 63 days after he left the nest did his wings grow to full length and no doubt the process of muscle toughening takes longer. Thus the attainment of competency in flying as well as hunting, for which flying ability is essential, cannot be hastened. Nature takes her course and sets the pace. Try as they will the parents are unable to shorten the training time any more than they can hurry the incubation period for eggs laid late in the season. The rate of physical and mental development of a young eagle cannot be forced any more than a human child's can. Life is just too short for many annual breeders and their young suffer from inadequate training when they get a late start in life.

CHAPTER THIRTEEN

Hunters of the skies

When I began to assemble material from my records for this chapter I felt confident that I had the full story of Black Eagle hunting methods and stratagem, including their execution of the coup de main in a swift "kill and carry" action which I had often observed.

Then it occurred to me that I had never seen a kill from the ideal position for an observer. Hitherto all my views of stoops had been from positions behind or to one side of the attacking eagle, whilst standing either higher or lower than the victim. Never had I seen the strike in every detail, and all the action before and after it, from directly in front of the oncoming eagle and lower than the prey, so that I could see the eagle as it struck, and the carry through, from underneath and without obstruction from tail or wing. It has, however, been a common experience for me to see my eagles low overhead carrying a hyrax dassie straight from the kill to their eyrie.

Next evening the three eagles were all out hunting together over the Still Bay cliffs and I stayed late, following their movements in the evening haze and fading light from vantage points on the cliffs which were reached by rough baboon paths through the macchia scrub at a snail's pace compared with the eagles'. I did eventually see them make a distant kill in the dusk but from quite the wrong position, which gained me nothing. All the way home I was depressed. I wished desperately now to obtain full confirmation of the results of earlier observation before it was too late. Next day was the 5th December, 1968 and the eagles would soon be hunting further afield, preparatory to leaving their nesting territory to settle Eaglet in a new area.

The chances were remote of seeing what I wanted in the rough and dangerous terrain of the forbidding Still Bay and Rondebosch Rivier cliff country where most of the adult eagles' hunting was now being done. On arrival home, I explained the reasons for my late return but, somehow, my confidence had returned. With bedtime came a mood of optimism, unreasoned but firm.

The very next morning at 9.15 o'clock, I was to see exactly what I was so anxious to observe, and from the best possible angle, in ideal conditions of light and wind for the spectators and the eagle actors who staged a perfect performance of one of the most spectacular and thrilling acts in their repertoire. The fact that this kill was made so conveniently must, I think, be regarded as a strange coincidence. These things do happen in one's experience among wildlife, as in human affairs. The occurrence was most providential and came in the nick of time to solve my difficulty of obtaining full and complete data for my investigation of Black Eagle hunting skills. I was certainly very lucky. Or is there another explanation?

This is what happened that morning. I went to inspect some work near our lower dam which lies in a wide basin about 1,000 yards east of the eagles' nesting cliff. The basin is open on its south-side where the water, flowing through the dam, tumbles over a barrage and cascades into the sea, forty feet below. This opening is a break of about 240 yards wide in the line of coastal cliffs. Headland crags form its sides near the sea and a small rocky ravine leaves a projecting cliff spur to the eastern crag.

I was standing on a terrace under this spur, the top of which was about fifty yards distant and forty feet higher than my head, when I saw a family of dassies sunning themselves as they lay straggled about the rocks on the ravine side of the spur. They were facing westwards, towards the eagles' nesting cliff, and watching me below. One half-grown dassie, weighing perhaps four pounds, moved and sat higher than the rest on the spur top in full view.

I turned towards Eagle Hill looking for the eagles but failed to see them. Then one by one they left their nesting cliff and began circling above it, gradually rising higher on the east wind's upcurrent and moving in my direction. Having covered half the distance between us, Eaglet hived off and dropped down to a high midway cliff perch, whilst his parents took an inland course parallel to the cliffs. When they reached the dam basin they ringed higher, keeping over 200 yards north of the spur where the dassies still sat watching them with unguarded curiosity.

Together the two adult eagles then did some aerial stunts. Whilst Ave kept up the gambade, so holding the dassies' attention, Mavi swept on rapidly eastwards, after coming through a tight loop, then immediately swung south to launch her attack is a very steep stoop which gave her incredible impetus. At a point 140 yards behind the dassies, she levelled out, just skimming the tops of the low bushes there as she wheeled sharply right and swooped westwards along the cliff edge. Then, avoiding the corner pole of a dangerous barbed wire and jackal-proof fence, which crosses the head of the ravine, and using a solitary low bush, seven yards to the rear of the young dassie, as cover, she sped straight along the spur as an arrow from the bow. In a flash she was over the gap with the dassie slung lifeless under her, like a float on a seaplane. For the first time in this hunt, she began making wing in regular beats, so gaining the up-thrust needed for bearing the extra weight of the dassie. Her feet had shot out with split second timing as she swept over the dassie on fully stretched, levelled wings in a swift, smooth power-glide.

How puny and feeble this human word "feet" for an eagle's! Eight knotted, iron clawed grapnel toes, almost as long as a man's fingers, had driven deeply into the prey with the impact's force; then grappled with vice-like strength to crush out life in a deadly grip. How amazing also the strength of her lightboned, sinewy frame and braced pectoral muscles, tough and resilient as nylon cord! The sudden drag of the gathered dassie would have pulled a man's arms out of their sockets at that speed, given weight for weight. Mavi's left foot talons gripped the dassie behind its left shoulder and her hind (1st) toe claw must have severed the spinal column in the strike, thus extinguishing life instantaneously. Her right foot clamped its mouth, ensuring against the danger of a bitten tarsus or thigh, the claws doubtless driven through eyes and brain. All this was accomplished at high speed in an action so graceful and smooth that the power and strength behind it were not apparent, as they certainly are when, for example, a barracuda snatches a victim in a surfacing leap.

An Arab saying goes: Three things come not back, the spoken word, the lost opportunity and the sped arrow. A Black Eagle's attack is silent and deadly sure and the speeding arrow ascends with the victim.

The warning cry of another dassie rasped out long after the victim had been snatched away. None of them had even noticed Mavi launch her attack. She carried the dassie directly over me and I could see exactly how she clutched it and its hind legs hanging limp.

Before she had crossed the gap, Ave had joined her, taking up his usual protective position when she carries, about seven feet behind and above her. As she dropped out of sight behind the ridge near the nesting cliff he circled round once, collecting Eaglet en route for the meal they had brought him, and after giving him a perfectly executed demonstration lesson in hunting technique.

No whisper of wind through trimmed quills nor rustle of feathers had warned those sun-basking dassies. The harsh but belated screech of a parent was the only sound that betrayed this drama of a young dassie's death.

Concealed now in their dark, dank-smelling, dung-floored cavern, the stricken family lay low all day, waiting for cover of dusk before stealthily emerging. When dassies emit their weird night screams are they haunted by nightmarish fears of flying dragons? It is as likely that a Giant Eagle Owl (*Bubo lacteus*), adept as the Black Eagle in the art of silent approach, has swooped down, under cover of darkness, and snatched away another victim. The prolific, plump, muttony hyrax dassies are preyed upon by bird and mammal predators, diurnal and nocturnal, but Black Eagles are their bane.

It must not be thought that Black Eagles have a set form of hunting or that the hunt just described should be labelled typical. They do have their own inimitable style, however, which can often be recognised in the patterns of their hunts designed to cover a wide variety of situations and conditions, opportunities and hazards. The hunt action described followed a fairly common pattern with conditions very favourable.

Black Eagles usually hunt in pairs, when domestic duties permit, or in threes when training young, and set about it methodically. Ave and Mavi may often be seen relaxed and apparently resting on their favourite rock perch east of the nesting cliff which commands a splendid view of the coastal cliffs and inland hill slopes. In fact, whilst storing up physical energy, they are using their powerful eyesight all the time to spy out the land and note the movements of dassies and other animals.

They will sometimes fly far out to sea or well inland, parallel with the line of cliffs, either to reconnoitre or to manoeuvre for an attacking position, having regard to the direction of the wind, which will be used for attaining speed in launching the attack; cover for the approach, which will ensure surprise; and, above all, a clear coast immediately beyond the point of capture, because, to strike and snatch away by binding to the prey in one motion appears always to be a tactical objective. Then too, a chosen victim near a cliff edge cannot escape by dashing forward.

The stricken dassie is not necessarily carried far. Ave, with his slim falcon build, is not designed for lugging a nine pound dassie up to the eyrie. He will drop down with it to the nearest convenient safe rock and there eat part of it, starting with the head and shoulders, breast and lungs, and carry the remainder to the nesting cliff.

Or, if their nestling is old enough to be left alone, he will sometimes go and fetch Mavi to do the carrying to the nest or the terrace used for feeding below the eyrie after the fledgling has flown.

During the incubation period Ave must do most of the hunting and the carrying because their eggs and downy chicks are never left unprotected from gull and raven robbers. In such times he kills and carries mainly young dassies, weighing up to about four pounds, which he can cope with even in gale force winds since, much more frequently than not, strong winter winds have to be contended with along these coastal cliffs. This all fits in with their life's design because Black Eagles feed their downy chicks almost wholly on young, spring-born dassie mutton, selecting the choice, firm, red meat. But, as we have seen, no part of the dassie is wasted because the parents consume every bit, including fur, skin and bones.

There is one exception to this rule, and it is a very important one indeed in this hunting context. The intestines of an old dassie are often removed and flung away for the Black-backed gulls to grab. There are two probable explanations for this and they are, in a sense, complementary. The first is precaution against disease.

I was informed by reliable observers that, shortly after my arrival in 1965, all the dassies in Robbe Berg Nature Reserve nearby had died of an unknown disease. The eagles may well be aware of the danger to themselves of disease contracted from old, and perhaps ailing, dassies' intestines; but not the gulls.

Now the gulls often follow the eagles when they hunt, hoping for the intestines, but they appear to be mobbing the eagles and the eagles cleverly turn this to their own advantage. A mock scrap takes place high in the air over a dassie habitat and the curious dassies love to watch the fun. They are accustomed to having gulls flying just over their heads all the time and are quite unafraid of them. They often see the eagles high overhead and, so long as they judge the distance safe, do not run for cover. The mock fighters shift position and lose altitude until an eagle, seeing an opportunity, can feign boredom and drop down out of sight behind a cliff unobserved. This eagle is now set to whip over a cliff-face or round a corner and pick off an unsuspecting dassie in a surprise attack from the rear. Had the dassie been able to count or had it not believed that there was safety in numbers, its days would not have been numbered because of a missing eagle that never misses. When the eagles have killed all the world knows it because the gulls shout it from the cliff tops.

Cape Ravens are used for the same manoeuvre although their display of aggression is often serious later in the season. Two pairs nest over a mile away from Eagle Hill, one on either side. They lay eggs in August or September, when the Black Eagles are normally busy feeding their fledgling, and the ravens are most hostile if the Black Eagles cross their territory then.

The eagles always have the situation under control and, whilst using the mobbing ravens as hunting stooges, sometimes have to exert authority. The spectacle of a fleeing raven driven down to sea level with a swooping eagle hot on its tail is convincing proof of eagle pre-occupation for the fascinated dassies. Meanwhile the other raven has had a fright too and the eagle on high is free to stoop for a kill.

Scrub veld interspersed with grassy or bare openings large enough to permit clean stoops is quite often hunted. Dassies taking cover in the scrub are startled by overhead swoops and captured when crossing the openings to reach their hiding places. The action for taking a victim off the ground is a "stoop and throw up", followed by normal carry-away whilst making wing in heavy beats. Team work is excellent in this form of hunting. One eagle keeps the dassie on the run and the other stoops.

At other times the eagles hunt casually and apparently without a plan, as a man might go for a stroll in the bushveld with a gun. Freelance or teamed, they may be seen quartering the cliffs, always flying low and fast and altering course, with the result that they suddenly appear round a corner or over a hump to take their prey by surprise. The male must of necessity do a lot of this lone hunting during the incubation period.

Black Eagles are never seen sitting and waiting ready for a dassie to appear before dropping down on it from directly above, as a Crowned Eagle does in the forest. Whilst fast action and surprise are employed over cliff ranges, they occasionally circle round and wind-hover or hold wind over dense low scrub country, spotting and following the movements of dassies, hares, guinea-fowl, mongooses, cats or lizards. Then they will drop from the sky, feet first like harriers do, opening their wings at the last moment to check their fall as they pounce upon the prey. The danger of damaging wing feathers on wiry bushes is very real in this type of hunting and, probably for this reason, adult Black Eagles do not often indulge in it. They are clumsy hunters in these conditions.

Precipices, krantzes, cliffs, kloofs, crags and koppies are their natural hunting ranges and they prefer swooping low along the tops, exploiting the air currents to the full.

Where the Black Eagles hunt, their natural enemies of old, the baboons, retire to sleep by night and sally forth during the daylight hours to maraud and forage. The eagles attack on sight wherever they encounter a troop above their eyrie or catch them at a disadvantage on a precipice.

Attacks on baboons are launched in high vertical stoops so powerful that the young ones are knocked over the abyss to their death or immediate destruction when the attackers swoop on spread-eagled to despatch the fallen. Adult baboons may also be caught off balance as they rush forward to push or pull adolescents away from the cliff edge; and stray babies are snatched away from distracted mothers in the ensuing confusion.

Baboon hunting has its dangers for men and dogs, leopards and eagles alike. A baboon at bay is cunning and dangerous. They lash out with their long arms and grasping hands or snap and tear with their canine jaws in reactions as sharp as the scorpion tail reflexes which their own hunting trains them to anticipate. But they are no match for the relentless Black Eagles.

The baboons make good their escape through the bushes to carry on with their plunder and wasteful destruction of crops, fruit and wildflowers and birds', bees' and rats' nests. In times of drought they will kill new-born lambs and dassies and other mammals. And they will not forget to return again for revenge on next year's eaglet, if they can.

Nothing on veld or farm is safe from baboons and this is the reason for the Black Eagles' punitive wars against them. Protection of their eggs and nestlings and young eagles learning to fly is the first objective.

The male Black Eagle is exceptionally bellicose and fearless when performing this duty. I have seen Ave, always the protector and defender of his family, at his magnificent best in single combat with troop leader baboons. A rude track passes along the cliff above his eyrie which is used by baboons traversing the nature reserve. The leader usually scouts ahead of his troop to signal them on when the coast is clear. Ave may be far away hunting but always keeping a keen watch on Eagle Hill. Immediately a baboon is sighted, he soars over, first to warn Mavi on the nest as he swoops by, then round the crescent cliff-face, brushing past the baboon in challenge, and at the cliff's end, spirals up and back high over the nest again.

From this position he makes his first stoop at the baboon, his aim being to drive it back from the cliff top, not over. If the baboon stands his ground in defiance, Ave swoops over him, just out of reach, and, throwing up steeply, flies a high pitch for a second stoop in the opposite direction. Now he hurls himself upon his foe, his wing shoulders thrust forward and wings swept back apeak, his scimitar-taloned toes bared, ready to rip or grip. For a moment the baboon hesitates on the brink, mustering his pride before impending fall to death or disgrace. Then Ave, almost upon him now, suddenly throws out his wings full stretch and, with that, the baboon claps his hands over his eyes and flings himself backwards into the bushes barking in terror as Ave swishes over him. The retreat follows quickly with Ave harassing the baboons into rout.

When Sixtyfive was one month old my friend Bill Miller photographed him in the nest and we saw Ave attack a baboon when it appeared on the nesting cliff. A photograph was obtained of the baboon shortly before the second stoop.

Earthbound creatures in a precarious position will not often stand before the show of power of a Black Eagle hurtling down on it. The effect of the eagle's shock tactics, ending with the sudden fear of being entrapped under the expanded wing is to throw the most stubborn opponent into confusion. Some animals, however, such as the Cape Wild Cat (*Felis libyca*) and the Lynx (*Felis caracal*), when defending their young, and particularly the mothers, will face death before loss under such onslaughts.

Black Eagles are not disposed to attack man but make a fine show of aggression when their nestlings are molested, employing forcefully their powers of speedy flight in a noisy air action which gives the image of invincibility, as when driving away their natural enemies. Remembering in contrast the silence and stealth of their action when capturing dassies and their aerobatic feats, we are able to appreciate their versatility as aviators.

The habit of Black Eagles and other birds of prey of spreading their wings over prey or enemy is one of their most successful psychological weapons. Most wild animals and birds have a primal instinctive fear of being surrounded, ambushed or enveloped by their enemies. The spreading of wings is symbolic of mantling which is the pose assumed by a victorious eagle over its victim. Baboons are very intelligent animals and must often see Black Eagles in this pose and know its meaning.

Protection of their dassie flocks is another probable reason for the eagles' feud against baboons. The apes may not succeed in killing many dassies but they often scare them into hiding. There is no doubt that Black Eagles prefer dassie mutton to primates' flesh but, having killed a baboon, they will not waste it.

The total extent of the hunting ranges used by a Black Eagle family is not easily ascertained when the observer must traverse long tracts of rugged cliff country on foot. The exercise is immediately simplified, however, when one knows with reasonable certainty that there is only one breeding pair resident in a large area and that one is watching the movements of that pair without the complication of stray intruders.

My eagles hunt over about fifteen miles of coastal cliffs, four of which lie east of their eyrie. In this section the cliff line is broken by valleys and beyond it lies the Robbe Berg promontory, nearly four miles long and cliff ringed, but for some unknown reason—possibly too many fishermen and holiday makers and gulls—the eagles are seldom there. To the west of the eyrie is the vast Still Bay cliff complex, with the Rondebosch Rivier kloof running into it, and beyond is sheer sea cliff. Inland (northwards) the eagles sometimes hunt over farm veld and valleys, containing riverine bush and a few krantzes, in a belt about two miles wide and running along the coast. They will avoid human settlements. This thirty square mile tract appears to be an optimum Black Eagle hunting area and all they require for the ten months of the breeding and training season.

It is interesting and important to note that Ave and Mavi are seldom seen hunting the dassies that live on their own nesting cliff and in the crags, shelves and scree below it. These dassies clearly demonstrate by their behaviour that they are aware of the truce.

There appear to be two reasons for this situation. The area will be the playground for the young eagle for about two months after it has left the nest and here it will learn the rudiments of hunting among dassies that are often careless of danger. Then I believe it possible that Nature has planned this arrangement in order to ensure that dassies are not exterminated by Black Eagles and other predators which the Black Eagles discourage. The dassie colony living under the protection of the eagles may therefore be regarded as their breeding stock. As they multiply and spread further afield those on the perimeter are culled by the eagles. In short, Black Eagles may, in a sense, be regarded as dassie farmers. A continual supply of their staple food is ensured by protecting a domestic flock.

Black Eagles may also conserve pigeons for their young. The local population of Rock Pigeons (*Columba guinea phaeonota*) is densest on Eagle Hill's cliffs where they are seldom preyed upon by peregrine or lanner. They ignore the Black Eagles and are often seen sitting a foot or two away from them round their nest. One or two pairs serve as faithful companions of the growing eaglets, entertaining them during long waits between meals. And, strangely, when the young eagles make their first fumbling grabs at young rock pigeons below the eyrie, the older pigeons condone the breach of trust.

The eagles range farthest in December and January when dassie supplies are getting scarcer nearer home and their juvenile is able to accompany them on long forays. This he (she) usually does for the early morning and evening raids but sometimes stays near the eyrie taking things easy at midday in the long (15 hours), hot midsummer days. Black Eagle hunting is extremely energetic and tiring for youngsters who are not as well developed physically or as experienced in hunting or practised in conserving energy as their parents are. My Black Eagles, which live at the coast, are generally absent for most of January and early February, provided they have laid their eggs not later than early June and their juvenile is trained and ready to fend for himself. It will be remembered that Eaglet had a damaged wing and the 1968 winter was unusually long and cold. Sixtyseven was born late in the 1967 nesting season. Departure of the family was therefore delayed until the second week of January in these two seasons.

Training for hunting

Training of juvenile Black Eagles in hunting requires a course of at least eleven weeks. The course consists of four stages. The length of each is given here on the basis of my observations for Eaglet's 1968 course which proceeded stage by stage pari passu with Sixtyfive's course and may be considered normal. The final stages of Sixtyseven's course had to be foreshortened and her training was inadequate when she left the nesting area. The results of such training are serious and may be disastrous.

We have noted how much Black Eagles rely on their flying abilities for their hunting. The progress of their young in acquiring hunting skills is governed largely by their powers of flight.

The duration of the first stage of hunting practice is about 40 days, starting with the day on which the fledgling Eaglet left the nest when 92 days old. It runs concurrently with flying practice mainly below and round about the nesting cliff.

The play-way and learn-to-do-it-yourself methods of education are employed. During this stage the juvenile is left on his own for long spells whilst the parents are busy hunting and bringing food to the terrace below the nesting cliff for him. Although he is kept well supplied, his appetite is voracious. He begins by attacking dead sticks, usually rough and twisted ones about eighteen inches long and half to one inch thick. He pounces on them, drags them along the ground with one foot and savages them with his beak. Later he flies to rock ledges below the terrace, following adult rock pigeons, and there he succeeds in grabbing a young pigeon hardly able to fly itself.

When he has made more pigeon kills he tries for big game, but with little success at first because he lacks savoir faire in his brushes with the nimble dassie. Careless young dassies may occasionally fall victim later on. The main thing is that he has tried on his own and found out for himself how difficult hunting can be and how much there is to learn. And so he has learnt wisdom.

During the second stage, which lasts for about ten days, he spends part of each day, weather permitting, accompanying his parents for flying practice, and hunting demonstration. Eaglet's last day in this class was the 5th December, 1968 when the dassie hunt, described earlier, took place. This young dassie seemed to offer an excellent opportunity for putting a beginner through his paces and Eaglet was probably keen to show his mettle. But his parents would be fully aware of the terrible danger from that barbed wire fence which obstructed the easiest approach for the final swoop. This illustrates the need for parental guidance in the big snares of life. A Black Eagle hunter in South Africa today must know how to take his fences and telegraph wires and hedge-hop highways and railroads and steer clear of aircraft.

Having started on the third stage of about twelve days, the tiro now assists his parents in hunting by active participation under supervision. His playing days are over. He is learning to live by living in earnest, to hunt the hard way.

The final stage of about a month, before the young bird leaves the nesting area, is a very important one because he must now become fully self-supporting. He spends much time gaining experience, largely by his own efforts and initiative, whilst his parents keep in the background. But they still live, and sometimes hunt together, as a family, all coming back to the nesting cliff in the evenings to sleep on their own night perches, and no doubt the parents ensure that their youngster does not suffer unduly from hunger during this time.

At this stage, when Eaglet was 157 days old, he took part with his parents in an attack on a small trekker party of baboons, which appeared above Eagle Hill's sea cliffs, and helped to repulse them with verve and valour. The poor baboons were frantic because they had a wayward child with them and had to rush forward and knock it between rocks in their efforts to save it from certain death. It has not been seen again.

I saw Eaglet capture a total of five small dassies in the nesting cliff area and he must have had other kills there and elsewhere but I never saw him try seriously for a big dassie. It will be remembered that the eagles were ranging far and going on long forays now. On several occasions I saw Eaglet alone coming back high and fast from the direction of the Tzitzikama Mountains.

The economic value of Black Eagles

It has not been possible for me to keep a systematic check on the number of dassies brought to the nest or kills made by my eagles during the ten months breeding period. My duties in the nature reserve are multifarious and include reports and accounting every week-end which often preclude visits to the eyrie.

Sample checks have been made, however, and my figures are based on these. They are rough and ready and should be taken as indicative only. During the nestling period of three months, when a large percentage of the prey consists of young dassies up to about four pounds in weight, kills average three per day or, say, 270 for this period. For the remaining seven months we may work on a figure of two per day including kills not brought to the eyrie but assumed because of indications such as full crops and blood on beaks. A round figure of 420 is obtained giving a grand total of 690 kills by a Black Eagle family in its nesting area during the breeding season. On the debit side is one game bird—a guinea-fowl picked up after a shoot.

The dassie menace to pastures and veld browsing has been mentioned, as has the fact that four dassies eat about as much as one sheep. In an area stocked to capacity a resident Black Eagle family therefore conserves grazing and browsing for 172 sheep. Could any farmer wish for better friends? And they cost him nothing in return for their help.

Where Black Eagles, and other predators whose natural prey is dassies, have been exterminated, administrations and farmers are forced to take costly measures to control the depredations of these vermin. Such measures will always be necessary where dassies live in thickets too dense for eagles and others to hunt. But measures leading to total extermination of dassies in open areas can be ill-advised.

I know of an area where this was done at a time when Black Eagles were on the increase and beginning to exercise effective natural control. Because all the dassies were killed the eagles had to resort to killing lambs in order to survive, so all the eagles were killed. The last state is always worse than the first when man interferes with the balance of nature. The history of the use of poisons by man for controlling pests offers most convincing proof of this danger. Natural controls cannot be bettered.

If every farmer understood this, Black Eagles would be regarded as one of Nature's most precious gifts to southern Africa.

The Black Eagle language

I am handicapped in the study of sounds from the nest because of an inadequacy of hearing due to gunshot concussion and taking quinine regularly for many years when it was believed to be a prophylactic for malaria; and I also find the noise of the sea under the eyrie cliffs very disconcerting.

Nevertheless, I have very firm views on these eagles' language in general. As we have already learnt from following the course of Eaglet's life, Black Eagles can be seen speaking to each other and, with a little patience, one is able to interpret the general trend of their conversation by their expressions and actions as well as by the movements of their beaks. Watch a human mother singing her child a lullaby, scolding it, encouraging it, being firm or angry with it; watch her in all her moods from alarm when danger threatens to joyful pleasure.

The Black Eagle language covers just the same range. We can see, but seldom hear, when adult eagles speak; although the juveniles can easily be heard. As a general rule Black Eagle adults speak silently when listened to by humans, however good their hearing. At some stage in their lives between the time of leaving the nesting area at the age of about six months and attaining maturity, they begin to speak on a wave-length outside the human auditory range. And they can use that wave-length over distances too great for vocal communication. But when excited they revert to the use of vocal chords which we can hear.

Their warning or alarm cries resemble yelps or barks of dogs, baboons, jackals or Jackal Buzzards, gulls and once I distinctly heard Ave imitate the croak of a raven. Their cries of joy resemble the trilling whistle of Cape Rock or Sentinel Rock Thrushes (*Monticola rupestris* and *explorator*). These are their neighbours in mountainous country and at the seaside.

The downy nestlings begin with chirps and chink-chinks;

then go on to series of cheeps; the older nestlings use chirrups and the juveniles cheerreeps with a metallic ring sometimes. All these can be plain or plaintive. The words may be repeated up to twenty times in monotone without intermission. The flying juveniles also emit shrill cheeping yelps less hoarse than those of their parents.

Black Eagles are not silent birds as men often suppose. They communicate with facility and pellucidly amongst themselves. Adults speak to one another only when necessary and this is infrequent because of their great understanding of each other.

The nestlings, on the other hand, voice their feelings freely and are not repressed. Black Eagles, being good educationalists, believe that eaglets should be heard as well as seen. They believe in auditory as well as visual aids and demonstrations for instruction; in self-expression and speech training for their young. The eaglets' vocal organs must be exercised and well-developed so that they can learn to use the special Black Eagle wave-length.

Because I am hard of hearing I have made special efforts in my Black Eagle language studies. And I have been rewarded by an important discovery. The Black Eagle language is not only a living one but adaptable to local environments through mimicry. The raven-like Black Eagle warning call was common in the Maluti Mountains where Black Eagles and Cape Ravens lived together in bygone days. Here I have heard it only once from Ave when ravens were being troublesome. Both Ave and Mavi use gull calls here. I never heard them in the mountains up country. Mention has been made of a pair of Red-winged Starlings nesting in the base of the eagles' nest and keeping Eaglet amused. One day when I opened the hide window, shortly before Eaglet left the nest, he greeted me with a Red-winged Starling song call.

The eagles' mimicry of local birds implies that they are members of the local bird community, sometimes using a common language even if that involves rubbing shoulders with some rapscallions. The important fact arising is that birds and animals do form communities and work out a *modus vivendi*. When is Man going to use a lingua franca for better understanding between the nations?

CHAPTER FOURTEEN

Time to go

Russet, my Red Setter companion of the wilds, has left us. He has passed beyond this wide range of ours but I still follow him armed only with memories. He died still doing his work for his fellow creatures after a very happy day.

We walked through the valley on a morning when the wildflowers were at their best, and he had his usual fun with the Cape Francolin Pheasants who live there. He picked up and savoured the scent of many hidden friends as he ranged free. Then we went through the forest along the stream, which flows at the foot of Eagle Hill, to Mermaid's Pool where he paddled in the cool, shaded brook.

We came back over the weir and in a yellowwood glade found Runi, the Grysbuck doe, stamping her feet and challenging Russet to catch her. She loved this game with her big brother, who had a rufous coat so like hers and a long flowing tail. He shared lunch with me in the eagle hide and had his siesta. An hour before sunset we checked the Redwing Francolin Partridge coveys on the grassy slopes and the waterfowl on the upper dams. Back home, he gave Daphne his usual warm greeting and ate a good supper. At ten o'clock that evening he died. His great heart, always so happy in its work and master's pleasure, had run down at last.

We buried him in a copse of sweet-blossom Salie trees among the wildflowers at noon next day. His body faces the rising sun and its rays will soon sift through silver trees to reach the flowers growing on his grave. Daphne and I and the five men with us were not the only ones present. Overhead in heaven's glow both Ave and Mavi were circling slowly round and round, silently watching us all the time.

In the hillside fynbos scrub only twenty yards away a pheasant called twice as we plucked purple Senecio daisies and dropped them fluttering over him. The voice of the wild was paying homage to a friend whose work and spirit of friendship will live on forever. It was strange—but only for some—because Cape Pheasants seldom call at midday. And earlier that morning Ave had circled over the copse many more times than he would normally do. It is strange and yet it is true. This reads like fiction but it is not. Did the eagles and pheasants and the others know? But how did they? Why should they? These secrets are Nature's still. I wrote these words on the night after Russet's death. They were my silent observations and thoughts at his graveside.

Russet's great secret was gentleness. Behind his rare gift of detecting the presence of friends was concealed a rarer knowledge of the ways of the wild.

We found the spirit of fun among wild creatures and shared it: the pealing laughter in wings and voices of rising pheasants and the gay mirth in merry legs of romping buck. We knew their moods and reached for their trust, all hearts at peace.

Russet's purpose in life was to serve his master well. Whether that master used him as predator or conservator was not his choice. He served with a striving that brought him further along his kind's allotted path to perfection. The diverse paths of every form of life lead on to the fulfilment of an ultimate purpose designed by the universal planner and as yet transcending all human thought. Although there can be no known way ahead for each species, man in his own striving has learnt to seek the guiding hand of Nature's Master. By the study of Nature herself more will be revealed to men. Through my own study of the workings of Nature I have found new meaning in the lively oracles of God which are the most precious thing this world affords.

Black Eagle visitors

The 16th December, 1968 was the only day on which I observed another Black Eagle family near Eagle Hill and our eagles were apparently expecting them.

When patrolling the western boundary of the nature reserve at 9 o'clock I was greeted by an excited Ave. He had been sitting on the west-facing cliff of Eagle Hill obviously on the look out for something. Twice he took high, fast flights along and beyond Rondebosch Rivier Valley and then returned to collect Mavi and Eaglet and together they flew far beyond the valley in a north-westerly direction. They were soon back over Eagle Hill and, whilst Mavi and Eaglet remained, Ave made another reconnaissance in a northerly direction. He returned more excited than ever and this time they all went north-east; then moved south to the sea cliffs beyond our eastern boundary.

There another Black Eagle family party of three approached them from the direction of Robbe Berg and I thought there was going to be a fight. The two families passed each other moving in opposite directions; then suddenly wheeled round and intermingled in a fine display of aerobatics. For the next fifteen minutes they had a high spirited time together, diving along the coastal cliffs and gambolling above the hills. Several times the two juveniles separated to fly together. Then they all disappeared towards Robbe Berg, moving playfully along. Meanwhile I had come first to the top of Eagle Hill and then to a high point in the east of the reserve to watch them.

At 11 o'clock the two families came into view again for a short time over a neighbouring farm in the east. Next day the visitors had gone. It is interesting to note that they made a wide detour round the reserve when they came and that the meeting was not inside our eagles' immediate nesting territory. Never before had I seen Black Eagles congregate at the coast. By a coincidence the 16th of December is the Day of the Covenant in South Africa and I wondered whether the young of these families were being encouraged in a friendship. If they are of opposite sexes an early companionship could ripen, leading to mating in four years' time when they attain maturity. This is the time of year Black Eagles begin congregating in Lesotho and I expected our eagles would follow their friends, who appeared to be migrating, but they remained here for four more weeks and were seen on most days coming and going from the eyrie on their hunting expeditions together or singly.

The wingspan test

On the 18th December I had an excellent opportunity to compare Eaglet's wingspan with his parents'. The three of them, flying abreast in close formation, passed directly overhead and some forty feet above me on Eagle Hill. Eaglet was between his parents and I estimated his spread from tip to tip to be about five and a half feet compared with Ave's six and Mavi's seven. Later, on the same day, I saw that his wings were as long as his tail when sitting, indicating that he had now grown into Black Eagle balanced line. Eaglet was five months or 154 days old on this day. When Sixtyseven was ten days younger her wingspan was, I estimated, about a foot longer. It is a great pity that bird books ignore wing-spread measurements which are so important for naturalists.

Although Eaglet was now grown up and flying well and hunting on his own, he was not yet ready to take the final test for winning his Black Eagle wings. After the three eagles passed over me they proceeded to Rondebosch Rivier Kloof, rising steadily on the quite strong sou'wester. Having reached the top of the kloof they spiralled higher and then the two old eagles broke off in turn, taking incredibly steep dives to speed along the Still Bay West cliff down to sea level. Eaglet did not follow them as he was obviously expected to do. Instead he dropped down and began hunting over the grassy hills north of the kloof. Mavi came back and took him up again but he still refused to make the devil dive and returned to the eyrie cliff while his parents went on to hunt beyond Still Bay.

On this day I noted that Eaglet's plumage, seen from underneath and including wings, was almost as dark as his parents'. His head was now distinctly lighter than his mantle, tending to light yellow, almost white, over the eyes.

Three days later Eaglet took part in the attack on baboons which was described in the chapter on hunting. This must have taken courage after his frightening experience of them soon after he left the nest. At one stage Eaglet was fond of sitting on the wooden roof of Hide I because it was four feet higher than the cliff rim and enabled him to watch his parents hunting in the Still Bay area. His parents discouraged this, I think because they regarded the hide as a man's nest and wanted to avoid violation of territory. Another reason may have been that Hide I was seldom used after Eaglet left the nest and the baboons came to it then, if Ave was far away.

Portraits of Eaglet

During his sixth month Eaglet grew in grace and stature and into a very handsome young Black Eagle.

The sun shone with favour on him, portraying him with consummate skill in morning and evening lights especially.

I saw the spotlight thrown on him as he surveyed his hunting ground from a lofty crag—an alert figure of physical energy in black and tan plumage with rich rufous mantle and gold-spun cap, all burnished with solar gold.

I saw the sun capture him in wild, free flight, emblazon him with flashing brands as the gay young beau sabreur cut a fine dash along the wild coast's marine parade.

I watched him soar and gather into orbit, flashing signals back to earth through those orient windows in his wings.

Black Eagles of all ages, free in their natural element, always enrich our study of them with the characteristic poetry of their kind's life and age.

To each his own

In late December and early January Eaglet did much of his hunting alone, with a marked preference for inland hunting. I think one reason for this was that the gulls mobbed him so much when he was alone along the coast. But he must also have been attacked by birds of prey when infringing their territorial rights inland and I saw an African Marsh Harrier give him a rough time when he crossed its vlei. Whereas the gulls molested him indiscriminately, he probably soon realised that these raptors were defending territory.

That lesson was to be driven home to him, ironically by his own father, when it was time for him to leave his own home. On the 10th January, 1969 I observed Ave and Eaglet sitting together on East Perch between 9 and 11 a.m., but was too far away to judge their moods. Soon after 11 o'clock Ave chased Eaglet a little out to sea beyond South Point and knocked him a few times out there. Then they returned together to the nesting cliff and sat peacefully preening themselves and resting until noon when I had to stop observation.

I believe Eaglet knew as well as I did what Ave had intended him to realise. Next day the eagles left their nesting territory, Mavi and Eaglet going ahead of Ave. In the previous year it was Mavi who slapped Sixtyseven on her last day at home. On both occasions the parents' physical action was quite deliberate and firm but unharmful and they quickly made amends by passing the midday hours happily and peacefully with the youngster after this unpleasant duty was over. Perhaps it was more a home-leaving ceremony than a duty.

Ave

The 10th of January, 1969 was the last time I saw the Black Eagle family together round their sea cliff home. Next day I saw Ave alone high over Rondebosch Rivier Kloof. Mavi had already left on the slow journey northwards with Eaglet and he would join them in the evening, after they had been given safe passage through other Black Eagle territory. No tercel would hinder a migrating mother and juvenile at this season.

From the top of Eagle Hill I bid them god-speed. I would miss them for a while but would soon be roaming again with my sons in the Formosa mountain country where hosts of golden wildflowers bloom in summertime and damsel and butterflies sport freely on mountain peaks. We shall see no Black Eagles there, unless we search the hinterland beyond, but I shall remember far-off Lesotho and anxiously await the eagles' safe return for the next breeding season.

I wondered where Sixtyfive was now and what had happened to the beautiful young ring-tail, Sixtyseven, with her grave handicap of an inadequate education for the hard life she would have to lead alone and far from home. If she survives, she will lose her ringed tail on maturity and may find a good mate.

For Eaglet the portents had been favourable after his happy birthday and good education at home. My thoughts turned to Ave again; to his good nature and his simple trust in friendly men.

Good-bye for now Black Eagles. Fly free again. Our skies are your skies, our cliffs your homes. Farewell, Ave, Avatar-born to be man's friend and claim our protection and fellowship.

Epilogue

It is June of 1970. Much wind has passed over Eagle Hill since I finished writing the story of my eagles a year ago. Ave has seen seventeen more moons wax and wane. But Mavi is no longer at his side on that lonely, lovely cliff.

They had their nest ready for eggs early in June a year ago. Their courtship and mating season, enthralling as ever, was drawing to a close. Mavi had begun fetching rush leaves from a pond and was busy lining the cradle for her eggs. In the midst of these preparations she suddenly vanished from her Eagle Hill home.

She did not return.

I was unable to find any trace of her. To this day I cannot tell how or where she died.

I could not decide at the time whether or not Ave knew what had happened to her. His behaviour suggested that he expected her back. For three weeks he brought fresh sprays to the nest daily and sat for hours on his perch, waiting, waiting, waiting.

At times he would patrol the empty skies, but he never went far from Eagle Hill. He gave himself little time for hunting and became thinner.

After the 21st June there was a remarkable change in his

demeanour. Quickly he became his old purposeful self. He soared off into the bush and brought back dry branches and huge spiky sticks to the nest. He piled these on top of the platform with the spikes sticking out all round.
A fortnight later, when he had finished this job, the nest resembled a porcupine on the defensive, with quills bristling. Ave had made sure that no other eagle or raven would want to use his nest that season.

Having locked up his home he was free to go away. On the 7th July, a cold, cloudy day, he headed north-east over Formosa Peak on the strong south-west wind. His destination may have been the Maluti Mountains.

Three days later he was back and greeted me from overhead. In the afternoon of the same day I saw him leave the eyrie cliff and fly northwards once more.

He dropped his feet in farewell as he passed over me.

Then, to my amazement, I saw another eagle on the look-out perch. I investigated and found a young, mature female Black Eagle sitting there. The yellow skin of her face and toes had a fresh complexion and her feathers the purple sheen which marks a young bird. I estimated her age at about seven years. She was a lovely thing and radiantly healthy. That night she slept on the night perch which all juveniles bred on Eagle Hill have used after leaving the nest.

I knew then that Eagle Hill was her birthplace. Without hesitation I christened her Marvel. Ave had apparently fetched her home. Then he left her in possession of his eyrie. It seldom happens that a Black Eagle voluntarily surrenders his eyrie during his lifetime and I wondered why Ave had done this and where he had gone.

The months passed pleasantly for Marvel in the security of her father's estate and the wildlife sanctuary where food was plentiful.

On the 10th April, 1970, I had a bad scare when a Black

Eagle was poisoned in the foothills about twelve miles from Eagle Hill. Ravens had been found eating a lamb which they had killed, and the poison was intended for them.

On examining this bird I found that it was a female, middle-aged and younger and smaller than Mavi. Her wing-span was six feet six inches.

Two days later three Black Eagles were seen soaring together over Eagle Hill and one of them, a young male, remained, first as Marvel's companion and then as her mate. He came as a stranger to me and is still rather wary. I named him the Unknown Prince and later Marvel's Prince.

During the first week of May I watched the young pair in courtship play. Their favourite game is to ring high, with Prince above Marvel. Every few minutes she breaks away and flies off at a tangent, whereupon he swoops down and brushes her from above, yelping excitedly. Then they continue their circling movement or begin a new series of "multi-V" dives.

The main nest has been put in order and the platform is higher than ever, as a result of Ave's porcupine additions which had to be flattened. Marvel has been photographed on the nest and when plucking and bringing green twigs for the incubation basin.

And so the young couple is settling in. Ave's Eagle Hill will doubtless be used by his family for hundreds of years more, man permitting.

The winter of Mavi's death was a time of bitter memory. Anger and despair ruled me.

But as time passed and I watched Ave's calm resignation during the long, lonely days that followed, and later his deed of high resolve in passing on his heritage to Marvel, so despair passed away. I knew that our cause was not lost. The ways of a wild eagle could still inspire a man, as they have done through the ages.

Appendix

The breeding season

Black Eagles normally breed once a year in autumn and winter and lay two eggs but raise only one eaglet. Breeding activity and family duties within the home territory take up over ten months of the year and include much more than the incubation period of about 45 days and the nestling period of 92 days. Re-occupation and defence of territory, nest repair and courtship take place before the eggs are laid and the juvenile is cared for at home and trained by the parents for up to three months after it has left the nest.

The main events of the 1968 breeding season are set out in chronological order below:

Event	*Date in* 1968
Re-occupation of territory by Ave..	6th February
Return of Mavi	18th February
Repair of nest and early courtship begins	12th March
Lining of nest with green twigs and courtship intimacy begins ..	20th May
First egg laid (at sunrise or earlier) Larger and marked	1st June

Second egg laid (at sunrise or earlier) Smaller and unmarked	4th June
Both eggs carefully examined ..	10th and 17th June
Only the second-laid, smaller egg now in nest (filmed)	15th July
Egg hatched at sunrise. Chick and eggshell filmed at 9.30 a.m. ..	18th July
Eaglet leaves nest at 3.15 p.m. (Photographed testing wings at 12.15 p.m.)	17th October
Family leaves territory; Eaglet forever	11th January, 1969

Breeding season summary

Repair of nest 12th March to 31st May (incl.)	81 days
Incubation period 1st June to 17th July (incl.)	47 days
Incubation time (2nd egg) 4th June to 17th July (incl.)	44 days
Nestling period 18th July sunrise to 17th October 3.15 p.m. (incl.) ..	92 days
Eaglet's education continued at eyrie 18th October to 10th January ..	85 days
Total breeding season 12th March to 10th January	10 months (approx.)

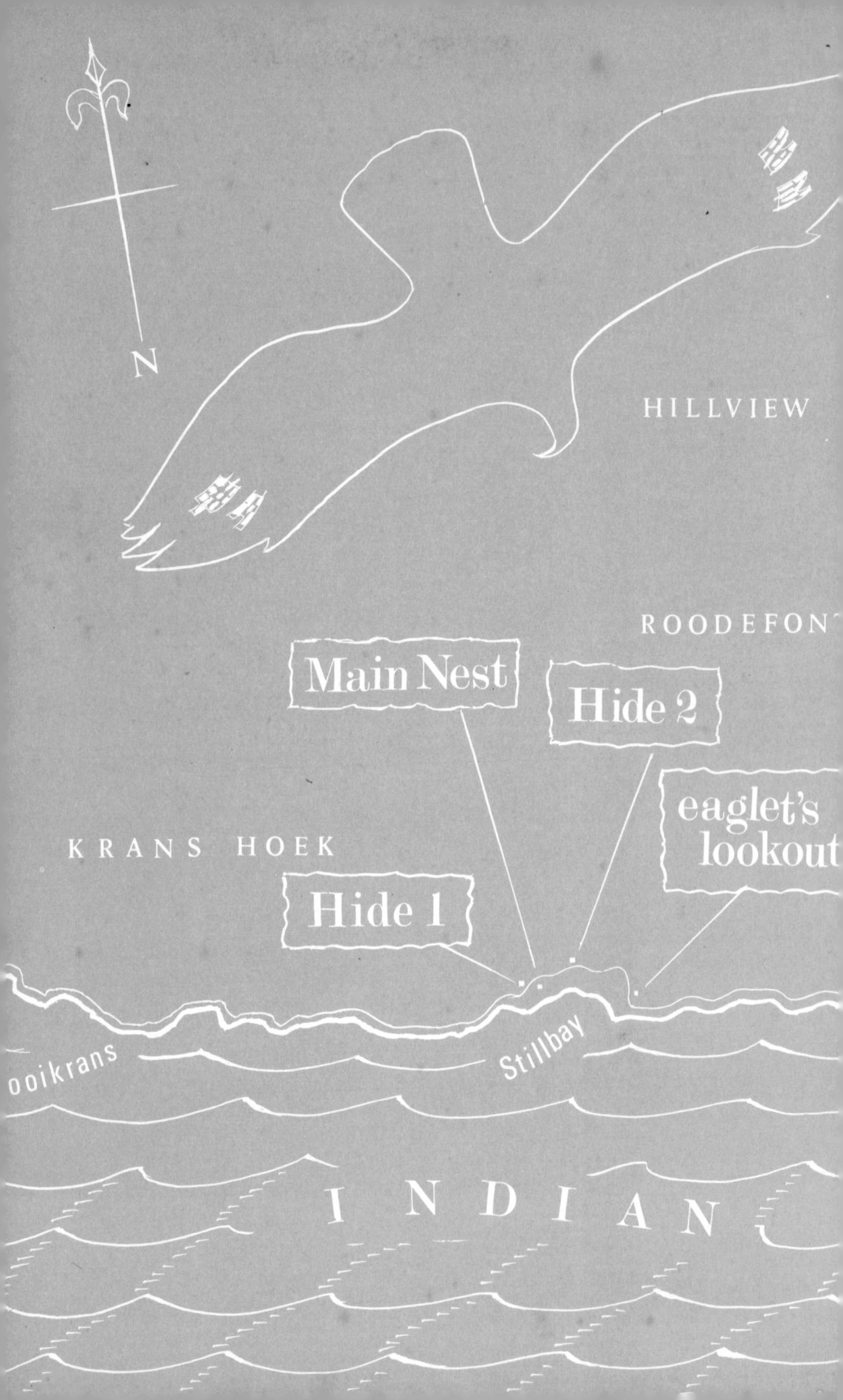
N
HILLVIEW
ROODEFONT
Main Nest
Hide 2
eaglet's
lookout
KRANS HOEK
Hide 1
ooikrans
Stillbay
INDIAN